The Magic of Physics

Can You Pull a Rabbit out of a Black Hole?

by
Llorrac Siwel
as told to
Prof. Richard Weiss

MACMILLAN

First published 1987

Published by
MACMILLAN EDUCATION LTD
Houndmills, Basingstoke, Hampshire RG21 2XS
and London
Companies and representatives
throughout the world

Typeset by TecSet Ltd, Wallington, Surrey
Printed in Hong Kong

British Library Cataloguing in Publication Data
Weiss, Richard
    The magic of physics: can you pull a rabbit
    out of a black hole?
    1. Physics
    I. Title
    530   QC21.2

ISBN 0–333–44278–4

Backword

Can one learn physics falling down a rabbit hole? Even with a firm hold on your metre stick and stopwatch, your own screaming will destroy your concentration and you'll have to rely on someone else to measure your acceleration due to gravity and your terminal velocity.

Ah! – but falling up a rabbit hole and out into the wonderland of nuclear magnetic resonance, non-destructive testing, laser weapons and orthogonal wave functions – that's real physics!

When Alice, the Mad Hatter and the Caterpillar are hired by Humpty Dumpty (alias Max Wells) to pursue physics on board the research vessel *Publish or Perish* the improbable thing happened – Max Wells' timely murder right between the covers of this book.

Is this just another crime? If you turn to the appendix and examine *20 Ways to Show You're a Physicist* you may recognize yourself. If so, you'll know the killer. If not, you've a lot to learn – you took a wrong turn up the rabbit hole. This book is required to pull you back by the hare.

Llorrac Siwel

To the Reader

If you hate physics and physicists, this book will woo you into loving both. If it fails then you should consider social work or needlepoint. No physicist worth his NaCl fails to exude the arrogance of that select community responsible for the atomic bomb, laser weapons and TV.

When Captain Max Wells, scion of that famous King's College physicist James Maxwell, the Clerk, sets up the Wells Institute for Physics Research he brings together on his private yacht *Publish or Perish* Alice, the Mad Hatter, the Caterpillar, and the famed Professors Schrodenberg and Seitell to do physics for fun and profit (mostly profit). Captain Wells' resemblance to Humpty Dumpty makes him both symbolic and dynamically unstable.

Alas! Captain Wells' timely death between the covers of this book creates both a mess and untold excitement. Even Lewis Carroll is called upon to examine the facts on his visit to the Daresbury synchrotron.

Does Prof. Schrodenberg spend all his time making his wave functions orthogonal?

Does Prof. Seitell fall in love with polymers?

Can Prof. Boltzius use laser holography to make a modern Stradivarius (to fiddle while London burns)?

Can Alice pass her A levels at only 13?

And what about spider silk and fledermaus soup?

These and other scientific questions remain unanswered in this romp through the Wonderland of Physics. If you don't enjoy the journey or fail to learn any physics you may be foolish enough to apply for a refund.

ACT ONE

SCENE ONE

(Alice and Professor Schrodenberg are walking in the Legendre Gardens. Alice picks up a white stone.)

Alice. And what does quantum mechanics tell us about this beautiful rock?

Schrodenberg. Not very much — not very much.

Alice. Even its colour?

Schrodenberg. Not even that.

(Alice picks up a fallen rose petal)

Alice. Surely something about this rose?

Schrodenberg. That's more difficult than the rock.

Alice. And this tree?

Schrodenberg. I believe it's oak.

Alice. But isn't that elementary?

Schrodenberg. L M N tree? Funny, I thought it was oak.

Alice. You told me that wave functions must be orthogonal — why is that?

Schrodenberg. Now *that's* elementary — if they weren't — why you'd disappear in a cloud of smoke. Remember, your ground state must be well-behaved.

Alice. Is that why people should not get excited?

Schrodenberg. A little perturbation is alright — but only if their ground state is still recognizable.

Alice. And all this was discovered in 1926?

Schrodenberg. Yes.

Alice. How did people manage before then?

Schrodenberg. The world was very confused.

Alice. But we *still* have wars.

(The two sit at a bench. The Professor unfolds a portable blackboard)

1

Schrodenberg. Let's start from the beginning — first, every electron has a wave function.

Alice. Is it worn like clothing?

Schrodenberg. Not quite.

Alice. How do we know?

Schrodenberg. Well to be perfectly truthful — we humans give it to them.

Alice. Perhaps they'd be better off without one.

Schrodenberg. Not if they're to be well-behaved. We must start with an atom. The electrons are all orthogonal — that is, their wave functions are.

Alice. But how do we know that?

Schrodenberg. Take helium — it's the simplest atom with more than one electron. If the wave functions are not orthogonal we get the wrong answers.

Alice. To what?

Schrodenberg. To the energies. If we make a gas of helium and excite the atoms in a gas tube, we can measure the light coming out — the light has very definite energies.

Alice. Are energies the same as wave functions?

Schrodenberg. No — no — each wave function has its own energy and we have a special name for it — eigenvalue.

Alice. And how do we find what that is?

Schrodenberg. We calculate it from the Hamiltonian.

Alice. Oh dear — this is confusing.

Schrodenberg. Remember — each electron has a man-made wave function from which we calculate the energy and we do that with the Hamiltonian.

Alice. What exactly is a Hamiltonian?

Schrodenberg. A simple device to look at the wave function and tell how much kinetic energy and how much potential energy the electron has.

Alice. Orthogonality — I like that word — six syllables. What about that?

Schrodenberg. That has to do with more than one electron. The wave function is like a wave — sometimes it's positive, sometimes negative. These positive and negative regions of one electron must just cancel those of the other electrons — that's orthogonality. If they didn't cancel they'd build up to a new wave function — but when they're orthogonal the electrons can live on the same atom and not disturb each other. It's like being a good neighbour.

2

Alice. I like that – orthogonality means being a well-behaved neighbour.

Schrodenberg. Shall we continue our walk?

Alice. Oh yes – I've had enough for now.

(The two rise and continue walking)

Schrodenberg. Before 1926 Niels Bohr tried to explain these definite energies in helium and hydrogen – but – but – not too successfully.

*(They pass the **Caterpillar** smoking a hookah)*

Caterpillar. Ohh – *(moans)*

Alice. Don't you know smoking is not good for you?

Caterpillar. Why? It was alright when Lewis Carroll was alive.

Alice. B – O – W – F.

Caterpillar. Bowf? What is that?

Alice. B – O – W – F; before orthogonal wave functions. Isn't that so Professor?

Schrodenberg. It would certainly make me sick.

Caterpillar. Orthogonality is one thing – but a good pipe is a smoke. I suppose you two have been at it again – Hamiltonians and all that nonsense.

Schrodenberg. It is not nonsense.

Caterpillar. It never got Disraeli elected – and look what happened to *Lady* Hamilton.

Schrodenberg. You have the wrong woman.

Caterpillar. So did Lord Nelson.

Schrodenberg. I see what you mean.

Caterpillar. Lewis Carroll said – Show me a Hamiltonian and I'll show you a confused student.

Schrodenberg. Very funny. It's not true once you learn it.

Caterpillar. By that time it's too late – you're in your third year and trying to find a job.

Schrodenberg. It's helped me – I couldn't live without it.

Caterpillar. You must be the only one for miles.

Alice. But Professor Schrodenberg is famous.

Caterpillar. Ha! Ask him to show you a real wave function – not one of those inferior man-made ones.

Schrodenberg. You've got me there.

(A white rabbit runs past)

Alice. *(Aside).* Where have I seen him before? *(To* **The Professor***)* Do you have a Hamiltonian to measure his kinetic energy?

Schrodenberg. No.

Caterpillar. You'd be better off with a stop-watch.

Schrodenberg. That's right. Besides it's hopeless unless you know his wave function.

(The **Mad Hatter** *comes by holding a scroll from which he is reading)*

Mad Hatter. *Recipe For Making An Atom* by Douglas Hartree
    First catch your electrons
    Give each an orthogonal wave function
    Unwrap your Hamiltonian
    Calculate the energy
    Keep the energy as low as possible —
    If necessary change the wave function to taste.

Schrodenberg. *(Applauds)*

Alice. Is this recipe well-tested?

Schrodenberg. 99 per cent accurate. What's one per cent between friends?

Caterpillar. So you know how to make atoms?

Alice. Any atom?

Schrodenberg. Any.

Alice. As many as you like?

Schrodenberg. Of course.

Alice. *(Picks up a piece of metal)* How about this piece of iron — there must be hundreds of atoms here.

Schrodenberg. Much more than that — in fact there are more atoms in that piece of iron than grains of sand at Eastbourne beach.

Caterpillar. Has Eastbourne changed that much?

Mad Hatter. It was alright when Mr Carroll went there.

Schrodenberg. I'm afraid quantum mechanics can't tell us a thing about that piece of iron — too many electrons.

Alice. Not a thing? So — how do we find out about iron?

Schrodenberg. You measure what you want to know.

Alice. But you said that helium . . .

Schrodenberg. Yes, one helium atom or even one iron atom but not a whole bunch of atoms living together in a solid.

Mad Hatter. Ha! The balloon just went up.

Caterpillar. At least that shows helium *can* be useful.

Schrodenberg. Just a moment — try a few atoms — like carbon dioxide

4

— one carbon atom, two oxygen atoms — that's one molecule and only 22 electrons. That we can calculate, although it's more difficult.

Alice. Why?

Schrodenberg. Three atoms — it's got a complicated shape — not a ball like one atom — more like three balls loosely stuck together.

Alice. So we can choose wave functions for electrons on atoms and small molecules but nothing more complicated?

Schrodenberg. More or less. Of course people try it on solids but it's hard to know how accurate they are — unless you've measured the answer beforehand.

Mad Hatter. You spend years in college learning quantum mechanics and it's all a lot of hot air.

Schrodenberg. As long as the air is just CO_2, N_2 and O_2 — but not if it's polluted with big molecules — then it's impossible.

Mad Hatter. If we had to wait for quantum mechanics to calculate the properties of steel in order to make automobiles — we'd still be on horseback.

Caterpillar. And would we have a pollution problem *(holds nose).*

Schrodenberg. Ah — here comes Professor Seitell — he should know about iron.

*(**Professor Seitell** carrying a copy of his latest book on Solid State Physics enters)*

Seitell. Good day.

Schrodenberg. Good day — Alice is curious about this piece of iron — can you calculate its density — I mean 99 per cent accurately.

Seitell. No.

Schrodenberg. Atomic arrangement?

Seitell. 99 per cent accurately?

Schrodenberg. Yes.

Seitell. No.

Schrodenberg. It's magnetism?

Seitell. No.

Alice. How about its colour?

Seitell. Not really.

Mad Hatter. Can it conduct?

Seitell. Brahms or electricity?

Mad Hatter. Ha-ha — electricity.

Seitell. I can't say how well.

Caterpillar. Is it affected by acids?

Seitell. Too difficult.

Alice. Then why do you teach Solid State Physics?

Seitell. To sell my books, of course.

Mad Hatter. And if I read your book — could I go far in my job?

Seitell. What do you do?

Mad Hatter. Design computers.

Seitell. Not unless you marry the boss's daughter.

Alice. Haven't solid state physicists accomplished anything?

Seitell. We discovered silicon.

Caterpillar. A new element! Oh good.

Seitell. No — an old one. It had been lost for decades. It was neither a conductor nor an insulator — we helped it decide.

Alice. What is it now?

Seitell. A semi-conductor — sometimes it conducts.

Alice. How do you use it?

Seitell. First, we cut it into thin slices.

Mad Hatter. A chip off the old block — ha, ha, ha.

Seitell. We add impurities — some technology and voilà! You have computers.

Alice. All this from your book?

Seitell. No — but it does sound more authoritative.

Alice. What else is in your book?

Seitell. How about magnetism?

Caterpillar. Very useful when you're lost in the woods.

Alice. Yes, I remember — there are three elements that are magnetic — iron, cobalt and nickel.

Seitell. Actually four — everyone forgets about gadolinium.

Mad Hatter. Four out of 106 — the chosen few.

Seitell. Professor Schrodenberg can help explain that.

Alice. Back to atoms again — we seem to owe everything to them.

Mad Hatter. They certainly keep us together.

Schrodenberg. Actually we're back to electrons.

Mad Hatter. Not those nasty wave functions?

Schrodenberg. Would you like something your maiden aunt understands?

Mad Hatter. Oh? — you've met her?

Schrodenberg. No thanks — I have my own. Did you know each electron is a magnet?

Alice. Why is that?

Schrodenberg. God made them that way — electrons are little balls of negative charge.

6

Caterpillar. Negative? Are you positive?

Schrodenberg. Absolutely — by scientific decree.

Mad Hatter. Amazing — you scientists can make electrons negative but you know nothing about iron — ironic!

Schrodenberg. And these little balls of charge are spinning around on their axes — just like the earth. Is that clear?

Mad Hatter. Like day and night.

Schrodenberg. That makes them magnetic — they never slow down. When we put electrons on atoms they sometimes have their north and south poles opposite so the magnetism cancels — but for almost all atoms they do not cancel and there is some magnetism left over. Do you think your maiden aunt would understand that?

Mad Hatter. Give her a cup of tea and you can tell her anything.

Alice. But didn't you just tell us only four elements are magnetic?

Seitell. That's when the atoms form solids. When they do, the electrons on atoms arrange themselves opposite to those on the next atom — the magnetism cancels.

Alice. Orthogonality?

Seitell. Precisely — good neighbours. But on iron, nickel, cobalt and gadolinium metals there are a few electrons that prefer to be on their own — they have their north poles pointing in the same direction.

Alice. Every piece of iron?

Seitell. Not exactly — if I squeeze iron very hard I can make a new form that's not magnetic.

Alice. *(Squeezes piece of iron)* Can I squeeze it in my hand?

Seitell. It has to be a million times harder than that.

Mad Hatter. Solids, solids, solids! Who can drink solids? I'd prefer a cup of tea.

(They all sit down for tea)

Caterpillar. Who'll be mother?

Alice. Allow me.

Mad Hatter. Why don't you write a book about liquid state physics?

Seitell. Too difficult.

Alice. Don't you know anything about liquids?

Seitell. Mostly intuition. For example, in solid iron every atom has eight near neighbours and they more or less remain neighbours. As you heat up a piece of iron each atom begins to vibrate a lot.

Mad Hatter. So would I if I sat on a hot stove.

Seitell. If you get the iron hot enough the atoms shake so much that they more easily move to a new neighbour and become liquid. Liquids can be poured because it's so easy for atoms to move about.

Alice. Why don't you mention it in your book?

Seitell. The equations are too difficult.

Caterpillar. Excellent reason! If you can't find an equation it's not important.

Seitell. Something like that.

Mad Hatter. How did the Scots ever learn to make whisky?

Seitell. They did it before my book was published.

Mad Hatter. Thank God for progress.

Caterpillar. Most of the world is liquid and we only study solid state physics.

Seitell. There *is* hydrodynamics.

Alice. Oh, dear — is that like Hamiltonians?

Seitell. Much more complicated — it's a study of how liquids move.

Mad Hatter. There's a whole science to that?

Seitell. Yes — very important.

Mad Hatter. If I study hydrodynamics will I be able to drink my tea faster?

Seitell. That depends — do equations make you thirsty?

Alice. They should — they're dry enough.

Caterpillar. Why study hydrodynamics?

Alice. All important things have long names.

Seitell. Like turbulence.

Alice. There you go again.

Seitell. Also very important.

Alice. How important?

Seitell. If I knew how to cure turbulence I'd be rich.

Caterpillar. Your book costs enough.

Seitell. Try the paperback.

Caterpillar. I mean that one — we caterpillars never use hard covers.

Alice. What is turbulence?

Seitell. Think of a ship moving through the water. At first the water moves smoothly past the hull — but the water molecules that hit the side of the ship bounce off and hit other water molecules. Suddenly the water molecules are going in all directions — that's turbulence.

Alice. Can you measure turbulence?

Seitell. We can tell at what point it starts by inserting a hot wire into the water. When the water flows smoothly (we call that laminar)

8

most molecules pass the wire — only a few hit the wire. When turbulent, more molecules strike the wire and cool it more rapidly. There's a sharp temperature drop.

Mad Hatter. Very clear — couldn't you grease the side of the ship?

Seitell. Perhaps — what happens when the grease wears off?

Alice. Does turbulence slow the boat down?

Seitell. Precisely — it also happens with airplanes.

Alice. An airplane flying through water — that *would* cause turbulence.

Seitell. No — no — I mean in the air.

Mad Hatter. Ah — do we also have air-state physics?

Seitell. Sort of — aerodynamics.

Caterpillar. Anything about biscuits in your book — they're solid?

Seitell. Nothing.

Caterpillar. Your book takes the fun out of life.

Alice. Can people be orthogonal?

Schrodenberg. I'm afraid not.

Mad Hatter. That's why they build fences.

Caterpillar. You can ignore someone — that's a bit like being orthogonal.

Mad Hatter. Ignore? I'd say ignorance gets you nowhere.

Alice. More tea anyone?

Caterpillar. *(Yawns)* I'm ready for a nap.

Mad Hatter. Trying to ignore me, heh?

(The **Mad Hatter** *gets on the table and starts throwing teacups, dishes, etc.)*

Seitell. You're mad . . .

Mad Hatter. You want to see a little turbulence? Try being orthogonal to that. *(Throws teacup at* **Seitell** *— it crashes to the ground)*

SCENE TWO

(The court at Old Bayleaf. Seated are **Alice,** *the* **Caterpillar, Knaves, Queen of Hearts** *and* **Professors Schrodenberg, Seitell** *and* **Boltzius.** *In the dock is the* **Mad Hatter.** *)*

Bailiff. All rise for the Lord Chief Justice.

(Enter the **King of Hearts** *followed by a* **Knave** *carrying a judicial wig on a pole. The* **King** *sits on the bench, removes his crown and dons the wig)*

9

King. Bailiff, read the charges.
Bailiff. Everyone be seated. Will the prisoner rise?
Mad Hatter. Which is it, up or down?
Bailiff. You are charged with excessive turbulence and needlessly increasing the entropy. How do you plead?
Mad Hatter. How am I charged — negatively or positively?

(Laughter)

King. Order — order. *(Raps gavel)*
Mad Hatter. I'll have a pint of lager.
King. Good idea — call the first witness.
Bailiff. I call Alice.
Alice. Oh, dear. *(Takes the stand)*
Bailiff. Raise your hand — do you swear?
Alice. Never.
King. We're not finished with you yet.

(The Queen rises and approaches the witness)

Queen. Are you familiar with the first law?
Alice. Of what?
Queen. Thermodynamics, of course. Energy cannot be created or destroyed.
Alice. Oh — of course.
Queen. Did you witness the defendant break the law?
Alice. He broke a few teacups.
Queen. A few? Bring in the evidence. *(A Knave brings in a tray of broken cups etc.)* If that isn't a misuse of entropy — what is?
Alice. But the tea was almost gone.
Queen. Admit it — you don't understand entropy.
Mad Hatter. Who does?
King. *(Aside to Queen)* What about the cup you broke yesterday.
Queen. That was coffee stupid, not entropy.
King. *(Aside to Queen)* Is my wig on straight?
Queen. Call Professor Boltzius. *(Boltzius takes the stand)*
Bailiff. Your name.
Boltzius. Professor Stefan Boltzius.
Queen. Are you an expert on entropy?
Boltzius. Don't you recognise the name?
Queen. Answer the question — don't tell me your problems.

Boltzius. A vital ingredient in Free Energy.

Mad Hatter. Free? I'll have some. *(Laughter)*

King. *(Raps gavel)* Order, order.

Mad Hatter. Where's my pint?

Queen. How do you view free energy?

Boltzius. With great fondness but sometimes with alarm.

Queen. Answer the question or heads will roll.

Boltzius. When you heat anything the atoms try to find the paths of greatest freedom — wouldn't you? The more freedom of movement the greater the entropy.

(The entire court rises and claps hands)

King. Brilliant, Professor!

Boltzius. Water has more freedom than ice and steam more than water. Water boils to increase the entropy.

King. I thought it boils to make tea.

Boltzius. Entropy.

King. Is that what broke the tea cups? I thought it was the Mad Hatter.

Queen. We must clear this up — I call Professor Seitell.

(Seitell takes the stand)

Bailiff. You've brought your own bible?

Seitell. I always carry my book with me.

Bailiff. Do you swear on the latest hardback revision of *Solid State Physics*?

Seitell. Absolutely.

Queen. *(Holds up broken cup)* Why did this cup break?

Seitell. I believe the Mad Hatter imparted excessive velocity to it.

Mad Hatter. I object — it was intact when it left my hand — after that it was on its own.

Seitell. It suffered brittle failure when it struck the ground, the excess stresses caused a small crack on the surface to move rapidly through the ceramic.

King. How rapidly?

Seitell. A thousand miles an hour — a crack moving in such hard materials pulls the atoms apart with great force. Of course, it doesn't happen much in metals.

Queen. Why?

Seitell. Dislocations.

11

Mad Hatter. Four syllables! It must be important.

Seitell. In a metal like iron the atoms are all the same and they can move past each other easily — one looks just like another — furthermore the dislocations help them. In a ceramic teacup there are several different kinds of atoms and they prefer remaining where they are. That's why it is hard but brittle.

Queen. And how do you measure entropy?

Seitell. Measure the specific heat, divide by the temperature and do your sums properly.

Mad Hatter. If I had to go through all that I'd give up tea.

Seitell. The whole basis of thermodynamics is free energy. For this you must know three things — the heat you put in, the temperature and the entropy.

King. Yes, yes, of course — but do you believe the Mad Hatter's guilty?

Seitell. Of course — he's obviously never read my book.

Mad Hatter. I object . . .

King. Very well, your witness.

(The **Mad Hatter** *steps down and approaches* **Professor Seitell***)*

Mad Hatter. Are you telling the court that if I read your book the teacup would not shatter?

Seitell. No.

Mad Hatter. When I threw the teacup at you — what did you do?

Seitell. I ducked, of course.

Mad Hatter. *(Slaps bench)* Why didn't you catch it?

King. Precisely! If you had caught the teacups you wouldn't be wasting the court's time.

Seitell. But — but — that's not cricket.

King. I doubt if you could bowl a maiden over. The prisoner is released.

(There is jubilation in court as the **Mad Hatter** *is hugged by* **Alice** *and the* **Caterpillar**. **Professor Boltzius** *shakes his hand)*

King. Bailiff — charge the new defendant.

Bailiff. Professor Seitell — you are charged with neglect — by permitting brittle failure to occur, by frequent revisions of your textbook, causing confusion and unnecessary entropy. How do you plead?

Seitell. But — I was working on my next revision.

Bailiff. Enter the dock. *(He enters)*

King. That was a rapid phase change, heh?

Queen. I call Professor Boltzius back to the stand.

(Boltzius takes the stand)

You have testified that Free Energy rules the world. Has Professor Seitell given this proper recognition in his latest revised edition?

Boltzius. To be honest, I don't believe so. After all, *every* substance tries to maximize its free energy.

Queen. Just as I thought. Do you see any reason for permitting Professor Seitell to keep his typewriter?

Boltzius. No — that will only produce another revision. You'd think by this time he would have got it right. He should also be required to read my book on thermodynamics.

(Professor Seitell rises and hurls his book at Professor Boltzius)

Seitell. *(Shouts)* That book is atrocious!

Boltzius. Ha! You didn't refer to it once in your horrible book!

King. Order — order.

Mad Hatter. I'll have *two* pints of lager.

King. Will the prisoner remain standing? Professor Seitell I pronounce sentence on you — death by elocution.

Queen. Stupid — you mean electrocution.

King. Close enough.

Seitell. *(Shouts)* I warn you, that won't work. My father was an excellent conductor.

King. Symphony or streetcar?

Seitell. Electrical — he was struck by lightning and survived.

King. Lock up the prisoner — bread, water, cold air — and a copy of Professor Boltzius's book.

Seitell. No! No! Not that book — it will surely kill me.

(The Knave drags out Seitell screaming)

King. Court is adjourned. *(Raps gavel)*

SCENE THREE

(Red-Ink Gaol. Alice visits Professor Seitell.)

Seitell. It's good of you to come — I'm going mad in solitary.

13

Alice. Have you read Professor Boltzius's book?

Seitell. What else? I've read all the graffiti 22 times. *(Looks around)* Have you heard the one about . . .

Alice. *(Coughs)*

Seitell. Oh – I'm sorry – I forgot.

Alice. I baked you a cake.

Seitell. That was kind.

Alice. *(Gives cake)* There's something special in it.

Seitell. Oh? What?

Alice. *(Whispers)* A copy of your book.

Seitell. Hardback or paperback?

Alice. Hard.

Seitell. Good – that means an extra £1.23½.

Alice. I read your chapter on electricity. You know – since they're going to . . .

Seitell. Very thoughtful of you.

Alice. Why do only certain things conduct – is it those nasty wave functions?

Seitell. I'm afraid so. In metals the valence electrons can be almost anywhere so they move easily through the metal, but in insulators there are gaps between the atoms where it is difficult to move.

Alice. Orthogonality?

Seitell. Partly.

Alice. And in semi-conductors?

Seitell. The valence electrons can move easily if you give them some help – either by heating or voltage.

Alice. But people – they aren't metals are they?

Seitell. No – there is a thin layer on our skins that is an insulator – also our bones. But there is so much salt in our body that a strong voltage causes conduction – a little like semi-conductors but more complicated. Actually, I wish we could talk about something else – you know . . .

Alice. Of course. Don't despair, we're working on your appeal.

Seitell. I do have a lot of that, don't I? It's my book.

Alice. We've found an expert witness. *(Guard appears)* I must go now.

Seitell. Thank you Alice for the . . . you know – the cake.

Alice. Don't eat it all – you might get indigestion. *(Whispers)* The third chapter.

Seitell. Yes, I'm working on that. *(Alice exits)*

Guard. Nice cake – but why is it shaped like that?

Seitell. Heh?

Guard. I've seen round cakes and square ones but why a rectangle?

Seitell. I see what you mean.

Guard. She hasn't done anything foolish like baking a saw into the cake?

Seitell. If there's a saw in that cake I'll eat my own book.

Guard. That's good enough for me. Back to your cell.

SCENE FOUR

(Newgate Court of Apples — formerly the Court of Appeals until computerised. Seated on the bench are the **Appeals Judges S. Ware** *and* **H. Ware** *who look like Tweedledum and Tweedledee and often speak in unison.)*

Judges. Bring forth the prisoner.

*(***Professor Seitell** *is led into the dock by the* **Knave of Diamonds** *and the* **Knave of Spades** *— the Solid State Contingent)*

Bailiff. You are guilty of overcharging. Is there any reason not to carry out the sentence?

Seitell. Is that reason enough to overcharge *me*?

Judges. Are you represented by counsel?

Alice. May I address the court?

Judges. How dare you — we are not envelopes.

Alice. We have secured distinguished counsel — Benjamin Franklin.

(The entire court rises and applauds)

Judges. But he is dead!

Alice. I ask the court not to mention that in his presence.

Judges. Very well, ask him to approach the bench.

Bailiff. Will the honourable Dr Benjamin Franklin enter the court?

*(***Franklin** *enters in 18th Century attire and carrying a kite. He approaches the bench)*

S. Ware. Are you qualified to act in the defendant's appeal?

Franklin. My son William completed his studies in the Middle Temple. He is a barrister.

H. Ware. Is that your bastard son?

Franklin. We all make mistakes, your honour.

S. Ware. Is he the one who became the . . . Royal Governor of New Jersey?

Franklin. Yes, your honour.

15

H. Ware. Not a bad mistake.

S. Ware. *(Points to kite)* Is that new evidence you have?

Franklin. Yes, your honour. May I explain? We are all familiar with the electric spark one can draw when combing the hair.

S. Ware. How would you know?

Franklin. *(Feels bald pate)* I have many friends who . . .

H. Ware. *(Winks)* Ha, Ha – we're both men of the world, I see.

Franklin. Could it be, I asked, that lightning was the same phenomenon – the same colour and all that. On a stormy day I observed, as my son William flew the kite.

S. Ware. The bastard son?

Franklin. When it comes to electricity I'm no fool. Sure enough the metal key at the end of the string permitted me to draw a spark to my knuckles. I had proved my thesis.

H. Ware. And for this discovery you became a fellow of the Royal Society.

Franklin. Yes, your honour.

S. Ware. I've always had my doubts about the Royal Society – this confirms them. Dr Franklin, what kind of string did you use on the kite?

Franklin. Silk, your honour.

S. Ware. Isn't silk an insulator? How could it conduct?

Franklin. It is possible to charge up an insulator by adding electrons – only a few and very slowly. Of course the string was wet, that helps.

H. Ware. Isn't water an insulator?

Franklin. You don't know Philadelphia water. Besides water fills the empty spaces in the silk thread – capillary action. The pressure of the air pushes the water into these empty spaces.

S. Ware. All very clear – but what about the defendant?

Franklin. I don't believe the Royal Society has read his book.

H. Ware. No – no – your new evidence.

Franklin. Oh yes – electrocution – the punishment does not fit the crime.

S. Ware. Why – have you read his book?

Franklin. I can't afford it, sir.

Judges. Just as we thought – did he not increase the entropy?

Franklin. Perhaps, a little and perhaps he overcharges but – 100,000 volts to his body – you'll run down his resistance.

Judges. You have a point. What alternative humane punishment do you suggest?

Franklin. Simple – change the 100,000 volts to 100,000 ohms.

16

Judges. *(Rap gavels)* Let justice be done.

(The shock wave of the gavels causes the Goddess of Justice to tip the scales precariously as her blindfold falls off. The entire court rises and applauds except **Professor Boltzius***)*

Boltzius. I object — I object.
Bailiff. Silence! Silence!
Judges. In addition, the defendant abandons his plans for a revised edition.

(The entire court bursts into wild pandemonium)

Boltzius. I don't object — I don't object.
Judges. Court recessed.

SCENE FIVE

(Red-Ink Gaol. Gathered around **Professor Seitell,** *who appears rather nervous, are* **Alice,** *the* **Mad Hatter,** *the* **Caterpillar, Professor Schrodenberg** *and* **Father Goldenestor***. The clock reads 15 minutes to midnight.)*

Alice. You must be brave.
Schrodenberg. Chin up and all that English stuff.
Seitell. Is it time?
Father. Soon, my son — are you prepared?
Seitell. Yes.

*(***Father Goldenestor** *opens his Bible to read)*

Seitell. No — not that — read this. *(Hands him a book)*
Father. But — but — this is your book on the *Theory of Solid State Physics* — with some chocolate icing on it.
Seitell. Read from Josephson — Chapter 7, verse 8.

*(***Father Goldenestor** *opens book)*

Father. *(Reads)* One of the most brilliant examples of solid state theory predicting an important experimental result and leading to a Nobel

17

Prize is the Josephson Effect which occurs in the layer between a metallic superconductor and its normally non-conducting oxide.

Seitell. Beautiful — beautiful.

Father. I've studied a little physics — what is a superconductor?

Alice. Should you? Just now — it's almost midnight — you know.

Seitell. Ordinarily, electrons moving in a metal encounter resistance from the vibrating atoms and they will bounce off them. The higher the temperature the more the atoms vibrate and block the electron paths. But in some metals at low temperatures pairs of electrons — with their north poles opposed — can move in perfect harmony with the vibrating atoms. All resistance ceases — hence a superconductor.

Father. Excellent — the pearly gates will surely open to greet you my son. Shall I continue?

Seitell. Yes please.

Father. *(Reads)* If a very thin oxide layer is sandwiched between two superconducting metals the electrons can find paths, sometimes called tunnels, between the superconductors. In order for this to happen the two superconductors are in precise phase with each other. A magnetic field can alter this, since the electrons are negatively charged. The Josephson Junction can be used in the most sensitive device to measure magnetic fields — sensitive to one part in 10^{10} of the earth's field.

Seitell. You read beautifully — I've always thought Solid State Theory is next to Godliness.

Mad Hatter. *(Aside)* Too bad it's so dirty.

Seitell. What is that?

Mad Hatter. I'm sorry. You're right. Cleanliness is next to Godliness.

(The church bell tolls midnight. The **Knave of Hearts** *enters carrying a small 100,000 ohm resistor on a silicon tray and places it before* **Professor Seitell.** *A dirge drifts in from the church where a midnight mass is being held.* **Professor Seitell** *picks up the resistor —* **Father Goldenestor** *makes the sign of the integral —* **Professor Seitell** *quickly swallows the resistor and faints)*

Schrodenberg. No — no you weren't supposed to swallow it.

Mad Hatter. What do you expect of a theoretician?

Caterpillar. What do we do now?

Alice. Rush him to a hospital.

18

(The scene shifts to Middlegender Hospital where **Professor Seitell** *still in a faint is on the ultrasonic scanner. The others watch as* **Dr Ray Leigh** *examines him)*

Leigh. I don't see anything — has he eaten recently?

Father. Yes — he thought it was his last meal — two orders of bubble and squeak, a pork pie, two pints of bitter, fish and chips, plum pudding and, of course, the 100,000 ohm resistor.

Leigh. And he appeared alright before he swallowed the resistor?

Alice. Oh, yes — a bit nervous.

Leigh. *(Shakes head)* The straw that broke the camel's back.

Father. I've studied a little physics — how does this machine work?

Leigh. The heart of it is a piezoelectric crystal.

Caterpillar. Piezoelectric — five syllables — very important.

Alice. What is it?

Leigh. In most crystals the atomic arrangement is identical if you turn them upside down. That's called centrosymmetric.

Mad Hatter. Centrosymmetric — five more.

Leigh. But quartz is non-centrosymmetric.

Mad Hatter. Quarts! I always order pints.

Alice. Non-centrosymmetric — gosh that's six.

Leigh. In quartz the atoms are displaced when viewed upside down so in an electric field the atoms move. With an alternating field the atoms vibrate one way then the other. If the quartz touches the skin it sends vibrations into the body.

Caterpillar. How fast do the atoms move?

Leigh. Well — humans can hear up to thousands of vibrations per second — this quartz crystal vibrates at about 10^{10} hertz.

Mad Hatter. If I shook that fast I'd hurt also.

Leigh. Not at all. The atoms in Professor Seitell's body are vibrating less than one-millionth of the distances between each other. The atomic vibrations from the temperature of his body are more like one hundredth of the distance between atoms.

Caterpillar. So we're shaking all the time.

Mad Hatter. Some more than others.

Caterpillar. Very funny.

Leigh. If the 100,000 ohm resistor is still in his stomach with all that other food it may be difficult to find. Perhaps we should wait for it to take its natural course.

Mad Hatter. With all that resistance won't it make him constipated?

19

Alice. That's not funny. *(To Doctor)* How do you see anything with this machine?

Leigh. Different parts of the body have different densities and so they shake differently. When there is such a difference the waves bounce back into the quartz crystal. The process is reversed and the quartz produces an electric signal that ends up on this picture tube.

Alice. Fascinating.

Leigh. If he hadn't eaten so much . . .

Alice. Could the resistor get stuck anywhere?

Leigh. It's hard to say — I'll authorize an NMR scan.

Father. What's NMR?

Leigh. Nuclear Magnetic Resonance.

Alice. Wow! Nine syllables — that beats everything.

Leigh. It's three separate words.

Alice. How disappointing.

(Professor Seitell, still in a faint, is wheeled into the NMR scanner room where Dr Felix Purr-Sell is waiting)

Mad Hatter. Holy smoke, what is that?

Purr-Sell. It's a huge magnet to put people into. It produces a field 10,000 times stronger than the earth.

Alice. Oh yes — the earth *is* a magnet. I wonder why?

Purr-Sell. No one seems to know.

Caterpillar. But there's iron and nickel at the centre of the earth.

Purr-Sell. Yes but it's probably not an ordinary magnet because of the high temperature and pressure. It may have to do with the liquid centre of the earth and the way it moves — but it's all a guess.

Alice. How do we know the centre is liquid?

Purr-Sell. By analyzing the shock waves from earthquakes — they travel differently in solids and liquids.

Alice. What are you doing now?

Purr-Sell. I'm strapping the patient on to this bed which can rotate him in all directions inside the magnet.

Mad Hatter. A few turns in that and the meal might come out the same end it went in. I'd rather not be around then.

Purr-Sell. What are we looking for?

Alice. A 100,000 ohm resistor.

Purr-Sell. Inside of him? Is he a robot?

Alice. No — that's Professor Seitell — he swallowed it.

Purr-Sell. Really? I didn't recognize him. I understand he's working on another revision of his book.

Alice. Yes – he was until the judge . . .

Purr-Sell. . . . and you're certain you want to save him?

Father. I studied a little physics – how does this work?

Purr-Sell. The nucleus of each hydrogen atom which is called a proton is a small magnet.

Alice. Just like an electron?

Purr-Sell. A thousand times smaller. When we place the Professor in the magnet the protons in his body try to point in the direction of the field, although the temperature vibrations shake them around so much that only a small percentage point towards the poles.

Caterpillar. Is this a Polish invention?

Purr-Sell. Why?

Caterpillar. I heard poles.

Purr-Sell. Magnetic ones – just like the earth's.

Mad Hatter. Stupid – you have the brain of a caterpillar.

Caterpillar. Thank God for little things.

Purr-Sell. Anyway, the small fraction that do point towards the poles will precess just like a spinning toy top when I add a small rotating magnetic field of just the right frequency.

Alice. What frequency is that?

Purr-Sell. Millions of hertz. We can tell when we are at the resonant frequency because the precessing protons rapidly lose their energy to the atomic vibrations. Also – the protons in the body vary a little because each has an electron near it. Remember – the electron is also a magnet a thousand times bigger. In different parts of the body the electron's influence alters the resonant frequency. We can focus our attention on special parts by varying the frequency.

Alice. Can you find the 100,000 ohm resistor?

Purr-Sell. But that's carbon – did it have a coating of any kind?

Schrodenberg. Some lacquer – that's got hydrogen.

Purr-Sell. Yes – but I'll need another resistor just like it so I know what frequency to use. Do you have another?

Alice. Gosh – no.

Purr-Sell. I don't think I can help – it will take too long to get one – the resistor may be eliminated by then. You might try Radiography.

(Seitell is beginning to revive as he is wheeled into radiography where they meet Dr Runt-Gen)

Seitell. *(Moans)*

21

Father. *(Makes the sign of the cross)* Blessed are the meek for they shall inherit the earth.

Runt-Gen. Father — I've studied a little Bible — could you explain that to me.

Father. Jesus taught that you must be like children to enter into heaven.

Mad Hatter. You're in the wrong pew if that sermon's for Professor Seitell.

Father. I'm only doing my job.

Caterpillar. Talk him out of his fourth revision — that would be merciful.

Mad Hatter. I thought the judge did that.

Runt-Gen. What's the problem here?

Alice. He's swallowed a carbon resistor.

Runt-Gen. I can't see carbon with X-rays — couldn't he swallow something heavier?

Alice. What about the lacquer coating?

Runt-Gen. Still too light — I need contrast-high atomic weight.

Caterpillar. Contrast? Compare Professor Seitell with Professor Boltzius.

Runt-Gen. No — no — *inside* Professor Seitell.

Alice. How about the two small metal leads on the resistor?

Runt-Gen. Now you're talking — why did he swallow the thing?

Mad Hatter. He couldn't catch a teacup. Straight at him, too.

Caterpillar. Tell us, doctor, we've all studied a little physics — what are you planning?

Runt-Gen. I'll look for the metal leads on the resistor — they're probably some alloy — maybe zinc, bismuth, lead, tin — all elements much heavier than those in the body.

Alice. But isn't there zinc and iron in the body?

Runt-Gen. In very low concentrations — parts per million — not concentrated as in the metal leads.

Seitell. *(Moans)*

Alice. Professor — you're going to be X-rayed, you swallowed the 100,000 ohm resistor.

Seitell. *(Revives)* Where am I? Is this heaven?

Mad Hatter. You were expecting miracles?

Alice. You're in Middlegender Hospital, in Radiography.

Seitell. Radiography — five syllables — they know I'm important.

Mad Hatter. What did you say about the meek Father? The only thing they'll inherit from Professor Seitell are his outdated editions.

Caterpillar. You're going to be X-rayed.

Seitell. No — no — not that. *(Covers his genitals)*

Mad Hatter. Send him back to Radiography.

Runt-Gen. This X-ray tube produces a point source of X-rays. Electrons are accelerated to about 150,000 volts and focussed on to a small piece of water-cooled tungsten. The electrons passing close to each tungsten nucleus produce X-rays.

Alice. Why is that?

Runt-Gen. Since the electron is negatively charged, it is caused to swerve by the strong positive charge of the tungsten nucleus. In order to conserve momentum it emits X-rays – they're called bremsstrahlung.

Seitell. *(Screams)* You're not going to do that to me – this isn't heaven – it's hell!

Runt-Gen. Would you like a sedative?

Seitell. Only if it stops the X-rays.

Runt-Gen. It won't do that but it will calm *our* nerves. Here take this.

Seitell. Not another 100,000 ohms?

Runt-Gen. Just take it – here's some water. *(Seitell takes pill and faints)*

Alice. He's fainted again – he can't take his pills.

Runt-Gen. Good. We'll strap him against the wall in front of the X-ray film.

(Professor Seitell is held up with two straps on his wrists and one around his waist. His head is bowed and he resembles the crucifixion)

Alice. He looks so peaceful now.

Runt-Gen. The X-rays passing through his body will be much more strongly absorbed by the metal leads of the resistor and this difference should appear on the film. Everyone out of the room, please. *(They all exit)*
When an X-ray hits a silver iodide molecule in the film it removes an electron and changes its chemical response to the developer. Ready? *(Presses button)* There – it will only be a minute. *(Runt-Gen leaves the room with the film)*

Alice. I hope he'll be alright now.

Mad Hatter. That depends on the X-ray. It's like the two Irishmen, Pat and Mike, adrift on an iceberg. Pat says "Look, Mike, we're saved – a ship." Mike asks "what's the ship's name?" Pat says "I can just barely make it out – Titanic." Try being orthogonal to an iceberg.

Schrodenberg. Seitell's been through a lot.

Mad Hatter. If he studied a little experimental physics – he might have caught the teacup.

23

(Dr Runt-Gen runs in with the negative)

Runt-Gen. The negative is negative − it must have been too small. Try Endoscopy.

Mad Hatter. Which end is that?

Runt-Gen. That's Dr Spiegelglass.

Mad Hatter. That should make things clear.

(They all wheel Seitell into Dr Spiegelglass's section)

S'glass. What do we have here?

Alice. It's Professor Seitell − something he swallowed. No one can find it.

S'glass. *(Looks at Seitell's chart)* Ultrasound, NMR, Radiography. What is it, the Hope Diamond?

Caterpillar. 100,000 ohms.

S'glass. Not one at a time?

Alice. It's only that big. *(Shows)*

S'glass. How long ago?

Alice. Hours.

Father. I know a little physics − what do you plan to do here?

S'glass. Endoscopy. We run a bundle of glass fibres down into his stomach and send laser light down. We can see very clearly through the fibres, because the light does not leave the fibres − it always bounces off the fibre walls because of the difference in the index of refraction. We'll have to give him some anaesthetic first. *(He administers the nitrous oxide and looks through the fibre)*

Father. I've heard about lasers.

S'glass. *Light Amplification by Stimulated Emission of Radiation.* If you excite atoms to a higher energy state which is metastable − I mean it sits awhile before returning to its lower state − it is possible to tickle the atom and cause it to return to its lower state sooner by shining light of the right energy on to the atom. This light doesn't actually take part in the process. The light emitted by the atom is exactly in phase with the light that stimulated the process. You get a chain reaction and all the light is emitted at once and in phase. Einstein discovered the process of Stimulated Emission. *(pause)* There's a special surface coating on the fibres called cladding that helps keep the light in. I see he's had a big meal.

Caterpillar. Enough to choke a horse.

S'glass. Looks like sausage.

24

Alice. Right.
S'glass. Potatoes?
Alice. Right.
S'glass. Plum pudding?
Alice. Right.
S'glass. Ah — there it is — take a look, Father.
Father. By God — you've found it — between the fish and the pork pie.
S'glass. I think it's small enough to pass through naturally — a few days in hospital and we'll examine his stool. I could remove it but I don't think it's dangerous.
Father. He won't be the first to recover after last rites.
S'glass. You're all so interested in physics — how'd you like to be my guests tomorrow at the big race. I have an interesting experiment with my horse.
Alice. Endoscopy?
Mad Hatter. I hope it isn't the end I'm thinking of.
S'glass. Will you come?
All. Super, great . . .
Mad Hatter. I can see the newspapers now — Professor Seitell Survives 100,000 ohms — Fourth Revision Dies.

SCENE SIX

(Magnesium Sulphate Downs, formerly Epsom Downs; **Professor Schrodenberg, Alice,** *the* **Mad Hatter,** *the* **Caterpillar, Father Goldenstor, Dr Spiegelglass** *and* **Melanie** *— blonde, in mink, young, etc., etc. They are in* **Dr Spiegelglass's** *private booth which contains considerable laboratory equipment.)*

Alice. *(Peering through binoculars)* What is the name of your horse?
S'glass. Reflection, he's No. 6 — green and white.
Father. Would you suggest a little wager? Not that I'm a gambling man, mind you.
Mad Hatter. Never been into the temple, heh?
S'glass. I'll let you know soon, Father.
Alice. Isn't it unusual for a doctor to race horses?
S'glass. I used to be a horse doctor.
Caterpillar. Why change?
S'glass. For one thing, mink is not cheap.
Mad Hatter. Neither are blondes.

25

Father. I can see that. I've studied a little physics Doctor, can you explain all this equipment to me?

S'glass. Can you keep a secret?

Mad Hatter. About your horse or your blonde?

S'glass. Under each of my horse's hooves I have mounted a small pressure transducer.

Alice. What is a pressure transducer?

S'glass. Inside the transducer are some electrical wires whose resistance changes as the horse presses down on them. The signals are radioed to this instrument and recorded on the oscilloscope. Look — Reflection is now walking for his warm-up. As each leg presses the ground the pressure is recorded. All four legs show the same pressure so none of his legs hurts him.

Mad Hatter. Holy smoke — that's clever.

S'glass. Reflection will soon take a warm-up trot. From the signal shapes I can tell if he'll run well.

Caterpillar. Is that legal?

S'glass. No one's told me it isn't. There he goes into his trot — yes — I'd say he's worth betting on today.

Father. If you'll excuse me — I'm carrying last week's collection money and I think I know what to do with it.

Mad Hatter. Here's a fiver for me.

Schrodenberg. This is more exciting than quantum mechanics — here's five for me.

S'glass. Before I place my bet I must calculate the odds on my new computer — it's a Granny Smith.

Melanie. I thought your Grandmother's name was Sarah?

S'glass. Watch the race, dear. It's very fast, optical fibres.

Melanie. Optical Fibres? I don't see that horse listed.

Alice. Why is glass transparent and metal opaque?

S'glass. Professor, you can probably answer that one.

Alice. I suppose it's the wave functions.

Schrodenberg. Yes, actually. In glass the electrons are confined to the regions between the silicon and oxygen atoms. Light, or photons, consists of an oscillating electric and magnetic field at right angles to each other but it's the electric field that causes the electrons to shake at the frequency of light. When photons enter the glass they are absorbed by the electrons but are re-emitted in the same direction a short time later. That's why light slows down in glass — it pauses each time it's absorbed and sent out again. It slows down about 50 per cent.

Mad Hatter. I'd really slow down if I had to go through glass.

Alice. And in metals?

Schrodenberg. There the electron wave functions permit the electrons to spread out more uniformly in the solid — that's why the electrons can move about and carry electricity. When light enters the metal the electrons absorb the photons as in glass but before the light is sent out again the energy is changed into heat and causes the atom to vibrate more. Hence the photon is lost in the metal.

Mad Hatter. I see, said the blind man.

Schrodenberg. You see — wave functions are a man's best friend.

(Father Goldenestor returns)

Father. All bets placed.

Melanie. Wave functions — I must have my hair permed.

Alice. *(Through binoculars)* They're nearing the starting gate.

S'glass. I must hurry with my calculations.

*(The door suddenly bursts open and a **Constable** enters)*

Constable. I have a warrant for your arrest, Dr Spiegelglass — *and* your confederates.

S'glass. What are the charges?

Constable. Illegal transmission of race information. I suggest you all come quietly.

Alice. They're off!

Constable. You're backing Reflection?

S'glass. Of course.

Constable. Disqualified.

Father. Damn! *(Crosses himself and looks skyward)* Forgive me.

*(**Melanie** faints into **Father Goldenestor's** arms as a news photographer flashes a picture)*

SCENE SEVEN

*(Old Bayleaf detention room. All are confined awaiting charges. **Father Goldenestor** is examining the Wells' Gazette with the compromising picture on the front page.)*

Father. Look at that headline. *(Reads)* Science and Religion join forces in race track swindle – and look at that caption under the picture – Priest Examines Furs for Clerical Robes. This will, will – wait till the Bishop sees this! And just listen to this . . . *(Reads)* "In an unusual display of teamwork a well-known Harlot Street doctor and several famous literary personages combined their talents to bilk the racing public. Bail has been set at £1 million." *(Throws down newspaper)* How could they?

S'glass. *(Picks up paper)* And to think Captain Wells and I were old wartime buddies.

Alice. I hope Mr Carroll doesn't hear about this.

Mad Hatter. How about my five pounds?

Melanie. *(Looks at paper)* That damn photographer got my wrong side!

Father. Where am I going to get £120 to return to my church?

Alice. *(To* Mad Hatter*)* It's all your fault.

Mad Hatter. My fault? The court didn't find *me* guilty – I didn't swallow that resistor.

Alice. The teacup! That's what I mean.

Caterpillar. What about me? I'll be back to eating leaves – no more McDonalds.

Schrodenberg. Does anyone have any money?

S'glass. 23 pounds.

Father. Are you joking?

Alice. Ten pence.

Mad Hatter. Me, too.

Caterpillar. Half a crown.

Melanie. I only use credit cards.

Schrodenberg. I have 14 pounds – let's see, 23 plus 14 plus . . . we still need 999,963 pounds.

Alice. We'll be here for ever.

Caterpillar. *(To* Spiegelglass*)* Do you still think it's not illegal?

S'glass. We haven't been convicted yet.

(Two Knaves, *the* Ace of Spades *and* Two of Spades, *appear with the* Bailiff*)*

Bailiff. OK, you're all free – you've been bailed out.

All. By whom?

Bailiff. Captain Max Wells.

(They all leave the detention room and enter the Bailiff's *office.* Captain Max Wells, *looking like Humpty Dumpty, is there)*

S'glass. Captain — what a surprise! May I introduce Father Goldenestor, Alice, Mr Hatter, Mr Caterpillar, Melanie and, of course, you've heard of Professor Schrodenberg.

Wells. Yes — some of us have met before. I must apologize for my newspaper — an over-zealous editor.

Father. So — you put up the million pounds?

Wells. It was nothing — I give that away quite often to my readers.

Father. I dare say a contribution like that to my parish . . .

Wells. I'm afraid I've had an ulterior motive in having you released. Your race-track scheme was brilliant.

Caterpillar. But illegal.

Wells. Brilliance comes first. Now I have an interesting plan that needs your talent in physics and computers.

Alice. Sounds exciting.

Wells. You can help me increase circulation of my newspaper.

Mad Hatter. I knew there was a rub . . .

Wells. Just answer one question — what's used during the day but hardly at night?

Alice. Shoes?

Wells. Yes — but no.

Caterpillar. Daylight.

Wells. Try again.

Mad Hatter. Bicycles.

Wells. Not the answer I want.

Schrodenberg. Lawn mowers.

Wells. Very funny.

Alice. Your turn, Melanie.

Melanie. But — *I'm* used at night.

Wells. Telephones! Just think — telephones. That's going to be the key to my circulation. Gentlemen and Ladies — could you build a system that will transmit and print my newspaper in every home by using the telephone in the early hours of the morning? We could get a special rate at night — you awaken and there's the newspaper.

S'glass. Brilliant — simple too.

Schrodenberg. No wonder you're the genius of Bleat Street.

Wells. You'll need to design a fast, cheap printer.

Father. I could include a daily sermon.

Alice. When can we get started?

Caterpillar. What about pictures?

S'glass. Easy — we use a dot matrix.

Wells. And don't forget — how to fold the newspaper.

Alice. Yes — of course.

Wells. Remember — the key is the telephone. Now, I'm very busy — if someone will turn me around I'll be off.

S'glass. Hold it — if we succeed — what are we paid?

Wells. £999,999 — you still owe me one pound.

(Some Lackeys enter and carry Captain Wells off in a sedan chair, sitting on a velvet cushion)

Caterpillar. Did he look like that during the war?

S'glass. No — actually he was the model for 007.

Mad Hatter. Drop the seven.

Alice. Very funny.

(Enter the Bailiff)

Bailiff. You're all free to go — until the trial. One million pounds — I wouldn't give ten pence for the whole lot of you. You have a visitor . . .

(They all exit and meet their visitor)

Alice. Mr Carroll — what are you doing here?

Carroll. Actually I came back to visit Daresbury, where I was born, when I read about your difficulties — but I see you're free now.

Mad Hatter. Double O rescued us . . .

Caterpillar. Daresbury, heh?

Carroll. Yes — I understand it's changed a bit. Would you care to come along?

Alice. Oh yes.

Father. May I join in?

Alice. This is Father Goldenestor — an amateur physicist.

Carroll. I was a cleric myself. You're certainly welcome. If we hurry we can catch the 10.30 out of Euston.

SCENE EIGHT

(The synchrotron at Daresbury. **Dr Chad Wick** *is showing* **Alice, Carroll, Mad Hatter, Caterpillar, Father Goldenestor, Professor Schrodenberg, Dr Spiegelglass** *and* **Melanie** *about.)*

30

Dr Wick. The synchrotron is like a photon factory — it produces the highest continuous density of photons found on earth.

Mad Hatter. You haven't seen Las Vegas at night.

Melanie. I have! That's where the Doctor and I . . .

S'glass. *(Coughs)*

Melanie. Are you in a draught, honey?

Wick. Electrons are accelerated in a large circle inside this magnet. The electrons emit photons to conserve momentum . . .

Alice. Like in bremsstrahlung?

Wick. Yes — the photons come out of this window where we set up our experiments.

Schrodenberg. What energies?

Wick. From a few volts to tens of thousands.

Schrodenberg. Plenty of X-rays?

Wick. Yes.

Alice. Radiography?

Wick. No, mostly diffraction. There are three ways to use X-rays. In radiography we look for density differences in the beam passing through a body — in therapy we use X-rays to kill unwanted cells in the body — in diffraction we examine X-rays that scatter from materials to see how the atoms are arranged.

Carroll. Mind-boggling — this is a real wonderland. My father was vicar of the church up on that hill when I was born in the parsonage. This was open farm land.

Alice. I was born years later — you were a maths don at Oxford.

Mad Hatter. Maths? You should see what these physicists are doing with your . . .

Schrodenberg. *(Coughs)*

Carroll. It's all so interesting but I must visit the old church — I haven't seen the stained glass window they've added in my honour. *(Exits)*

All. Good-bye — it was nice to see you.

Wick. Would you like to see the rest . . .

(Messenger enters)

Messenger. An urgent telegram for Dr Spiegelglass.

S'glass. *(Opens telegram)* We must return to London at once — Professor Seitell has tried to commit suicide.

31

SCENE NINE

(Middlegender Hospital — enter Dr Spiegelglass and the others)

Nurse. Dr Spiegelglass, you're just in time. *(They enter Professor Seitell's room — he is in a coma)* He left this note.

S'glass. *(Opens and reads)* "When I realized that I am forbidden to publish my fourth revision I decided I could not go on. I had added at least 20 more pages of equations. My third revision published two years ago, Makemillion Press, 32 pounds hard cover, 18 pounds paperback, shall be my final *magnum opus*. Goodbye cruel world." Signed Professor Seitell FGH, IJK, LMN, RIP.

Nurse. And then he ate the entire manuscript.

S'glass. Oh, God — that will jam up his system.

Mad Hatter. That's a case of terminal indigestion.

S'glass. Quick, nurse, endoscope and tweezers.

Nurse. Here Doctor.

S'glass. *(Inserts)* I'll have to pull this out a page at a time — but this saves a major operation. *(Looks through eyepiece)*

Alice. Do you see it?

S'glass. Yes — nurse — ready with the waste container. Here we go — Free Electron Theory.

Schrodenberg. Yes — an early attempt to explain electrical conductivity by assuming some electrons are free to roam the metal at will — too crude to be useful.

S'glass. Let's get rid of that! Ah — Debye Theory of Specific Heat.

Schrodenberg. Yes — a simplified approach to the way atoms vibrate in a crystal — useful but unrealistic.

S'glass. We won't save that either. What do we have here — they look like necks and bellies. Ah! Fermi Surfaces — nice pictures.

Schrodenberg. More useful — it discusses the energy and momentum of the electrons that conduct electricity.

S'glass. Shall we save it?

Schrodenberg. Well — alright, although it still smells. *(Holds nose)*

S'glass. We're coming to a lot of equations now, but I can't find the chapter heading — save it?

Mad Hatter. 'Mathematics is the contraception to understanding' — *Confucius* — into the waste basket.

S'glass. He's beginning to come around. *(Removes endoscope)*

Seitell. *(Moans)*

S'glass. Professor — Professor — can you hear me?

32

Seitell. Where am I — still in hell?

(Nurse leaves with waste basket)

S'glass. No — no — you're in hospital.
Seitell. What happened?
S'glass. You tried committing suicide.
Mad Hatter. You blew it again — can't you get anything right?
Alice. That's very unkind. Professor, how are you feeling?
Seitell. I — I — I — the last thing I remember was . . .
Nurse. *(Returns)* You were eating your manuscript and asking for mustard.
Mad Hatter. Mustard? No! Some grains of salt — yes!
Seitell. I — I — remember. Why didn't you let me die?
Alice. But — such a horrible death! And then to be buried with your own manuscript inside of you.
Caterpillar. The evil that men do is oft interr'd with their bones.
Alice. I don't think you've got that right.
S'glass. We've saved some of your manuscript.
Seitell. You have? Oh, thank you.
S'glass. Here. *(Gives to Seitell)*
Seitell. Two pages? But — but . . .
Alice. You can always start again.
Seitell. You heard the judge.
S'glass. We'll appeal — we've found a distinguished expert.
Seitell. You think there's a chance?
Alice. Yes — yes — you must fight.
Mad Hatter. And learn to play cricket.
Seitell. You've given me a new life — how can I ever thank you?

(Enter someone looking like Albert Einstein)

One-Stone. Is there a Professor Seitell here?
Alice. Yes — that's the Professor.
One-Stone. I am Professor Halibut One-Stone. I'm to appear on your behalf.
Seitell. Not the famous — but I thought you were dead?
One-Stone. You can't kill a legend.
Seitell. This is the happiest day of my life.
One-Stone. We must start on your appeal at once. What do we have to begin?

S'glass. *(Gives the two pages of the manuscript)* Here.

One-Stone. *(Holding nose)* Fermi surfaces? Phew — it's not much — but
— not everyone is an Einstein.

Melanie. Who's Einstein?

(Curtain)

ACT TWO

SCENE ONE

(Newgate Court of Apples — Alice, Mad Hatter, Caterpillar, Professor Schrodenberg, Professor One-Stone, Melanie, Dr Spiegelglass *and* Professor Seitell *in* the dock.)

Bailiff. Hear ye, hear ye — the Honourable Judges Ware presiding. All rise.

(The Judges Ware *enter and take their positions on the bench)*

Be seated.

Judges. On whose behalf is this appeal?

Seitell. Me — Professor Seitell.

Judges. What? You again — you're trying our patience.

Melanie. Why, what did Patience do?

Judges. Who said that?

S'glass. Please forgive the outburst — it won't happen again. *(To* Melanie) Be quiet!

S. Ware. Do you have new evidence?

One-Stone. May I approach the bench?

H. Ware. Your name?

One-Stone. Halibut One-Stone.

S. Ware. Are you the *famous* One-Stone?

One-Stone. Does $E = Mc^2$? I represent the defendant in his appeal against the court's decision forbidding new revisions of his book.

H. Ware. So you question our decision? Tell us — is the speed of light constant?

One-Stone. Of course.

S. Ware. So are our decisions.

H. Ware. Is energy conserved in a moving frame of reference?

One-Stone. Of course.

S. Ware. Then save your energy.

One-Stone. But — but — you've just stated the entire basis for Einstein's equation — $E = Mc^2$.

Judges. We may be judges but we're not stupid.

One-Stone. B . . . but what about Professor Seitell?

S. Ware. You're right — he *is* stupid.

One-Stone. I'm chagrined.

35

H. Ware. Didn't you say you were One-Stone?

One-Stone. I beg the court to consider my appeal.

Judges. Very well — no one can say we're not tolerant.

One-Stone. *(Takes position before the bench)* I, the famous One-Stone bow before the learned judges' astute understanding of relativity. It frequently requires 20 to 30 pages of text and equations to present what the learned judges — Soft Ware and Hard Ware — have so eloquently pointed out to this court.

Judges. This One-Stone is no fool.

One-Stone. $E = Mc^2$ provides the useful basis for calculating the energy from nuclear reactors — we are all aware of that. But we should not overlook the gems in Professor Seitell's book.

Mad Hatter. *(Coughs violently)*

Alice. *(Slaps him hard on his back)* What's the matter?

Mad Hatter. I'm choking on those gems.

One-Stone. Let me quote — "If we construct the planes normal to the lines connecting the near and next near neighbour atoms in crystals we produce a figure commonly called the Wigner–Seitz cell — useful in building up the charge density . . ."

Judges. *(Snoring)*

One-Stone. Your Honours!

Judges. *(Still snoring)*

One-Stone. *(Much louder)* Your Honours!

S. Ware. *(Awakens)* I'll bid 4 no-trump.

H. Ware. I'll double.

One-Stone. The gems — your Honours — may I continue?

Judges. Continue.

One-Stone. "Band theory can predict the density of electron states in metals. It has been the subject of many approaches such as Orthogonalized Plane Waves, Augmented Plane Waves, Tight Binding, Free Electrons and many others."

Judges. *(Snoring)*

One-Stone. *(Coughs)*

Judges. *(Still snoring)*

One-Stone. *(Coughs more loudly)*

S. Ware. *(Awakens)* Do something about that cough, One-Stone, or you'll never last the day out.

One-Stone. May I continue?

H. Ware. *(Raps gavel)* Brilliant — brilliant — will the defendant approach the bench? *(Aside)* I can't take any more of this. *(Seitell approaches the bench)*

Judges. Professor Seitell — because of the brilliant presentation of Professor One-Stone we grant your appeal to proceed with the fourth revision of your book. *(Spectators break into dancing and general pandemonium)* Silence! *(Rap gavels)* Silence! However, you must reduce your book to half the number of pages.

Seitell. But — but — but — that's impossible. *(Turns green and faints)*

One-Stone. Thank you, your Honours, your wisdom is boundless.

Judges. Boundless? We are all aware of the curvature of space.

Melanie. Curvature? Are they talking about me?

SCENE TWO

(Reception room for visitors to Eye-See-Eye, the Great Industrial Plant. Present are Dr Spiegelglass, Melanie, Alice, Mad Hatter, Caterpillar *and* Professor Seitell.*)*

Receptionist. May I help you?

S'glass. We're here to see Dr Polly Murr.

Receptionist. Synthetic fibres?

S'glass. Yes.

Receptionist. Do you have an appointment?

S'glass. Yes.

Receptionist. Please wait. *(They sit)*

S'glass. I haven't seen cousin Polly for ages — she's just the one to help us on the design of the home printer for Captain Wells.

Alice. *(To* Seitell*)* Don't look so discouraged — we're sure you can do it.

Seitell. How — How? I've agonized over every word in that book.

Mad Hatter. With half the words you'll only have half the agony.

Seitell. *(To* Mad Hatter*)* The next time I get my hands on a teacup, you'd better not be around.

Mad Hatter. I can handle your best pitches.

Melanie. You should have seen some of the pitches thrown at me.

S'glass. Why didn't Father Goldenestor come along?

Alice. He's been called to see the Bishop.

(Enter Dr Murr)

S'glass. Polly! Thanks for seeing us — this is Alice, Mr Hatter, Mr Caterpillar, Professor Seitell and Melanie.

Polly. Come into my office.

(They proceed down the corridor and into a large office) What's this all about?

S'glass. We've got a hot project — we need some advice on the construction of an inexpensive but tough printer — something that won't crack when dropped.

Polly. You want a polymer?

Alice. We've all studied a little physics — what is a polymer?

Polly. Any long chain of repeating molecules — for example, suppose you have a cyclohexane molecule — six carbon atoms in a honeycomb ring each with two hydrogen atoms joined to it. It's a liquid. If the ring of carbon atoms is opened up and you add additional molecules you can form a chain of carbon atoms each with two hydrogen atoms — that's polyethylene — it's solid because the chains are so heavy — tens of thousands of molecules long.

Seitell. Remarkable — I've always wanted to know what a polymer was.

Polly. Polyethylene is man-made, but there are thousands of natural polymers like cellulose in trees, fingernails, hair, bones, celery, asparagus — nature discovered how to use long chain molecules, strength in one direction and flexibility in the other direction.

Seitell. Magnificent!

Polly. In polyethylene the carbon atoms are strongly bonded to each other and less strongly to the hydrogens. The chains stick to each other rather weakly so they bend easily. Nylon is a well-known example of a strong man-made polymer.

Seitell. Beautiful.

Alice. *(To Seitell)* Yes — polymers are beautiful.

Seitell. *(Glassy-eyed)* I mean Polly.

S'glass. What about the printer?

Polly. I'd recommend fibreglass in polyester — the fibreglass is strong and cheap and the polyester forms the matrix holding the fibres together — it's called a composite — cheap, tough, easily moulded.

Seitell. Beautiful — beautiful.

Polly. Actually, composites are an old idea — centuries ago in India huts were made of a straw in a cow dung matrix.

Mad Hatter. *(Holds nose)* Can you imagine what they were like on a rainy day?

Polly. Does Professor Seitell always do that?

S'glass. You mean stand on his head? No — I haven't done that myself since I was sixteen.

Polly. He looks cute that way.

Alice. What are some other composites?

Polly. Carbon fibres in epoxy — entire aircraft frames are made of it — strong and light — also tennis rackets.

Alice. Tell me more about nylon.

Polly. You start from a liquid called a monomer which consists of single nylon-type molecules, add something that polymerizes it — draw it through a die to make fibres and then weave it into fabric from which stockings can be made — simple!

Alice. Nylon is strong so I guess the die must be.

Polly. Interesting — because we've had a problem with the dies wearing out, even ones made from diamonds.

S'glass. How do you get a diamond that big?

Melanie. I know how. *(Shows diamond ring)*

S'glass. Not now Melanie.

Polly. You don't need a big one. Very small artificial diamonds are made at high temperature and pressure. These are sintered together into a die by using cobalt metal as a binder.

Alice. So the cobalt glues the diamonds together.

Polly. Yes, but only because the diamond dissolves a little cobalt. Well — these dies were wearing out too rapidly so we tried ion implantation — high-speed nitrogen atoms are shot into the surface of the dies and these have increased their lifetime tenfold.

Alice. Why?

Polly. Only a guess, but we think the nitrogen atoms prevent the diamond from cleaving. How long can Professor Seitell stand on his head?

Alice. This is the first time I've seen it.

Polly. What is his field?

Alice. He writes books on Solid State Theory.

Polly. Oh? He's *that* Professor Seitell — cute.

Alice. He's beginning work on revising his book.

Polly. From the bottom to the top? Look, would you all like to see my laboratory?

All. Oh, yes.

Polly. I've arranged for my assistant to take you around. I'd like to speak to Professor Seitell.

All. Great. *(They exit with the assistant)*

Polly. *(Stands on head)* Professor, could I speak to you?

Seitell. Beautiful.

Polly. I mean right side up — I was going to offer you some tea when the others returned. *(They both return to their feet)* Please have a seat. *(They sit)* You're revising your book?

Seitell. Book? Book? What book?

Polly. *Theory of Solid State Physics.*

Seitell. Oh — you're writing one also?

Polly. No — no — *your* book — aren't you revising it?

Seitell. Oh — that can wait.

Polly. I was wondering why you don't mention polymers.

Seitell. Very good idea — beautiful.

Polly. There's a lot of good theory.

Seitell. Oh? Very good idea — beautiful.

Polly. Where do you teach?

Seitell. Teach? Oh, yes, teach — London, beautiful.

Polly. Do you live in town?

Seitell. California.

Polly. Interesting. You teach in London and live in California. How do you manage?

Seitell. Long week-ends.

Polly. I should think so. *(Very long pause)* And composites.

Seitell. Very useful — very useful.

Polly. I mean in your book.

Seitell. Oh? What page is that on?

Polly. No — no — why not include it?

Seitell. Yes, yes — good idea.

Polly. And Ion Implantation — very exciting.

Seitell. Anything you say — beautiful idea.

Polly. *(Pause)* Perhaps I could help you.

Seitell. Could you? Really? Help me?

Polly. Professor?

Seitell. Yes?

Polly. You're standing on your head again — and here come the others. *(They enter)*

Alice. Very interesting tour.

Caterpillar. Couldn't you get Prof. back on his feet?

Polly. I thought I did — but he must think we're superconducting pairs — Professor — we're going to have tea now. Please — back on your feet. *(To the others)* We've just agreed to work together on the revision of his book.

All. What a great idea.

Alice. Where will you work on it?

Polly. Somewhere between London and California.

Mad Hatter. That sounds like Liquid State to me.

Alice. Very funny.

40

Polly. There's the tea now. *(Tea is served)*

Alice. I'll be mother. *(She pours a cup for* **Polly,** *then for* **Professor Seitell** *who is mooning over* **Polly** *and drops the cup which hits the floor but does not break)*

Mad Hatter. It didn't break.

Polly. Yes, a new ceramic — almost indestructible.

Alice. But how?

Polly. Very small grain size — cracks don't start.

Alice. Too bad we didn't have them several weeks ago.

Seitell. Beautiful — remarkable. *(* **Seitell** *starts throwing teacups at the* **Mad Hatter** *who ducks —* **Seitell** *laughs uncontrollably)* Beautiful — remarkable — ha, ha, ha.

SCENE THREE

(The See of Londinium. **Sister Theresa** *enters the* **Bishop's** *office.)*

Sister. Father Goldenestor is here.

Bishop. Send him in. *(* **Sister** *leaves,* **Father** *enters)*

Father. You sent for me?

Bishop. Please sit down. *(He sits. The* **Bishop** *takes out the incriminating issue of 'Captain Wells' Gazette'. He examines the picture on the front page, then stares at* **Father Goldenestor)**

Father. I — I – I . . .

Bishop. Yes?

Father. I — I can explain.

Bishop. No need — this is not a confessional. So you're interested in physics?

Father. Why yes — but — but . . .

Bishop. Did you know I studied a little physics at King's College in London?

Father. No.

Bishop. Yes — that's how I developed my interest in Divinity.

Father. I didn't know that.

Bishop. The Divinity Department is just over the Physics Laboratories.

Father. Really?

Bishop. At least they got the hierarchy right.

Father. *(Forced smile)*

Bishop. I was in the Physics lab. doing this Compton scattering experiment for Dr Sewering — that's the experiment where an X-ray makes

41

a collision with a single electron and you determine the electron's momentum from a measurement of the energy and momentum of the X-ray. It's the best method to tell you about the wave function.

Father. I see.

Bishop. Well — you know how students are — we caught a mouse and decided to do Compton scattering on it. We tied it up and left it in the X-ray beam over the week-end.

Father. What happened?

Bishop. It killed the mouse but not the bacteria in it. They never did succeed in getting rid of the smell.

Father. *(Forced smile)*

Bishop. I was told to move upstairs to Divinity where they specialize in death — not much chance to get into trouble there, you know.

Father. Interesting.

Bishop. Now — about *Captain Wells' Gazette*. I think I can overlook the picture on the front page — but £120 of the parish's money — tsk, tsk, tsk.

Father. But the horse won! If only it . . .

Bishop. . . . wasn't disqualified?

Father. You knew?

Bishop. I read the sports pages once in a while.

Father. I see.

Bishop. How long will it take to replace it?

Father. I'm hoping in a month or so.

Bishop. Alright — you have two months — but in the meantime keep your hands clean — understand?

Father. Yes, sir . . .

Bishop. That is all. *(Father Goldenestor exits)* Ha — ha — ha — ha — wait till the boys in Rome see this — ha — ha — ha.

SCENE FOUR

(The court at Old Bayleaf — lawyers' briefing room. Present are **Dr Spiegelglass, Melanie, Captain Wells, Father Goldenestor, Alice, Mad Hatter, Caterpillar, Professor Schrodenberg** *and the* **Famous Barrister Lord John Bury-Moore** *who looks like the White Knight.)*

Bury-Moore. Because of the technical nature of this trial we must be ready as witnesses. Now, above all, do not appear professorial — that will surely put the judge to sleep. Furthermore, you must not seem nervous or uncertain. If scientists only learned to make their talks simple, not use too many slides – if possible none at all – they would be less nervous. The only thing worth saying is that which your audience will remember for at least a few days and that's not slides with curves or numbers on it. Remember that. Captain Wells is not paying me £10,000 per day merely because I'm Lord John Bury-Moore.

Mad Hatter. *(Aside)* I'll bet it helps.

Bury-Moore. I have attended scientific conferences in my professional capacity as Legal Advisor to the Royal Egocentric Society. Conferences are merely a guise to give scientists free trips and enable them to catch up on their sleep during the lectures. Any questions?

S'glass. What approach is the prosecution likely to take?

Bury-Moore. They will try to discredit all of you – it was regrettable that *Captain Wells' Gazette* published that picture. Be on the alert for character assassination. If in doubt keep silent. If you're in trouble, I'll come to your rescue. Anything else?

Melanie. Will you call me as a witness?

Bury-Moore. Not if I can help it. Also cover your cleavage.

S'glass. Have you determined whether my experiment was illegal?

Bury-Moore. A moot point – I still don't know – I'll try to avoid that one.

Wells. I have the utmost confidence in Lord Bury-Moore. I hope in a few days you'll be back working on the newspaper transmitter. *(Bell rings)* Let's go – and good luck. *(They all enter the court; Spiegelglass goes into the dock. The King of Hearts enters as before)*

Bailiff. All be seated. *(Queen of Hearts approaches bench)*

Queen. I call Mr Hatter to the stand. *(Mad Hatter takes stand)*

King. So – it's you again?

Queen. Were you in the dock on an entropy charge exactly one month ago?

Mad Hatter. Yes – but I was innocent.

Queen. Just answer the questions. Were you called in to help the Spiegelglass gang obtain free energy?

Bury-Moore. I object.

King. *(To Queen)* Lord Bury-Moore objects, dear.

Queen. No further questions. I call Melanie Bayha.

S'glass. *(As Melanie passes Dr Spiegelglass – to Melanie, whispers)* Try to keep your mouth shut. *(She takes stand)*

Queen. Were you ever employed in Las Vegas?

Melanie. *(Nods)*

Queen. Are you a photographer?

Melanie. *(Shakes head)*

Queen. Then why were you arrested in Las Vegas for over-exposure?

Melanie. *(Shrugs shoulders)*

Queen. Did you join the Spiegelglass gang to obtain free energy?

Melanie. *(Shakes head)*

Queen. So — you're not free?

Melanie. *(Nods, then shakes head)*

Bury-Moore. I object.

King. Lord Bury-Moore objects, dear.

Queen. I call Father Goldenestor. *(He takes stand)* Is this your picture in the *Wells' Gazette*?

Father. Yes, but I can explain.

Queen. On the day in question were you found with £120 of betting tickets on Dr Spiegelglass's horse?

Father. Yes, but I can explain.

Queen. Were you not called in to see the Bishop two days later?

Father. Yes, but I can explain.

Queen. Was the £120 your own money?

Father. No, but I can explain.

Bury-Moore. I object.

King. Lord Bury-Moore objects, dear.

Queen. No further questions — I call Alice. *(Alice takes stand)*

Queen. How old are you?

Alice. Thirteen, but to many I'm ageless.

Queen. Did you obtain permission from Mr Carroll to join the Spiegelglass gang?

Alice. No — but I was only studying physics.

Queen. Did you not know you must be 21 to wager a bet?

Alice. No.

Bury-Moore. I object.

King. Lord Bury . . .

Queen. . . . -Moore objects dear. I call Mr Caterpillar. *(Caterpillar takes the stand smoking his hookah)* How long have you been hooked on that?

Caterpillar. Seven years.

Queen. You joined the Spiegelglass gang to pay for your habit.

Caterpillar. I . . . I . . . I.

Bury-Moore. I object.

Queen. *(Looks at King)* I know! I know! I call Captain Wells. *(Wells is lifted to the stand on his velvet cushion)* Did you post £1 million bail for the Spiegelglass gang?

Wells. Yes.

Queen. *(Shows newspaper)* Did your paper publish this article?

Wells. Yes.

Queen. Can I assume your paper made a mistake? *(Pregnant pause)*
*(*Queen *turns to* Bury-Moore*)* Aren't you going to object?

Bury-Moore. Not at all — I'd like to know the answer myself.

Wells. *(Coughs)* It was published on our science page as a brilliant
achievement.

Queen. Is that why your paper's headlines said . . .

Bury-Moore. I object.

King. Try not to upset Lord Bury-Moore dear.

Queen. *(Aside)* Wait till I get him home. *(To court)* I call Professor
Schrodenberg. *(*Schrodenberg *takes the stand).* Are you a Professor
of Physics at Gottingberg University.

Schrodenberg. I am.

Queen. Did you not place a £5 bet on Revelation to come in first
position.

Schrodenberg. I did.

Queen. And you expected him to have maximum momentum at the
time.

Schrodenberg. Oh yes.

Queen. Are you not aware of the uncertainty principle? — that you cannot determine position and momentum at the same time.

Schrodenberg. *(Gasps)* I never thought of that.

Queen. *(To* Bury-Moore*)* Aren't you going to object?

Bury-Moore. I didn't think of it either.

King. See here, Professor, we don't allow witnesses who defy principles
— you're dismissed.

Queen. I call Dr Spiegelglass — boss of the gang. *(Much excitement in
court)* Your name?

S'glass. Dr Spiegelglass of Harlot Street.

Queen. Will you tell the court the principle of the pressure transducer —
if indeed your gang has any principles left.

S'glass. It is based on the fact that when a metal undergoes a strain its
electrical resistance increases.

Queen. Explain that to the court.

S'glass. The strain disturbs the crystalline arrangement and the electrons
are more likely to run into the strained atoms.

Queen. So — you had electrons running as well as your horse?

S'glass. Yes.

Queen. Were the electrons officially entered into the race?
S'glass. No.
Queen. Double deception! The prosecution rests!
King. Lord Bury-Moore — is the defence ready?
Bury-Moore. Yes, your Honour.

(Bury-Moore rises, takes centre stage, waits for the house lights to dim and the spotlight to hit him)

The quality of mercy is not strain'd, it droppeth as the gentle rain from heaven. *(Bury-Moore bows as the entire court applauds)*
King. Brilliant! Brilliant! *(Bangs gavel)* Case dismissed against the Spiegelglass gang.

(Pandemonium breaks out — Captain Wells peels out one hundred £100 notes and gives them to Lord Bury-Moore. Suddenly we hear an uncontrollable laugh from the rear of the court as teacups fly everywhere, bouncing off the walls intact. Professor Seitell has appeared)

Seitell. Beautiful, beautiful — ha, ha, ha, ha — beautiful.
Wells. Bring that man to me!

SCENE FIVE

(Head-Ink Town Hall, headquarters of the Wells Conglomerate. Seated in the reception area are Professor Seitell and Dr Murr.)

Receptionist. Captain Wells is prepared to see you now.

(The two are led into his office where Captain Wells is surrounded by Lackeys carrying shields to protect him)

Wells. Good morning — you don't have any teacups, I hope.
Seitell. No sir!
Murr. I must apologize for the Professor — he hasn't been himself lately.
Wells. And who are you?
Murr. Dr Murr, Eye-See-Eye. Professor Seitell and I are collaborators.
Wells. Very well — now, I want to say I was very impressed with that demonstration of unbreakable teacups. Marvellous.

46

Murr. They were developed by Eye-See-Eye.

Wells. Very good. You see, a man in my position can't be too careful. I've decided I need a bullet-proof vest — something light-weight.

Murr. You want kevlar.

Wells. What is that?

Murr. A light-weight polymer of unusual strength woven into a fabric. It's used to make sails on racing boats.

Wells. It's a plastic?

Murr. Oh, yes.

Wells. And it will stop a metal bullet — how is that possible?

Murr. For lead ammunition, yes. Kevlar has a high breaking strength and absorbs the energy of the bullet. The soft lead bullet will deform, of course.

Wells. Interesting — I'd like one in egg-shell blue.

Murr. You can have any colour as long as it's yellow.

Wells. Is that supposed to be a yolk?

Murr. No sir.

Wells. Does the army use kevlar?

Murr. Sometimes — but if the bullets are high velocity armour piercing, one needs a ceramic vest — that's much heavier.

Wells. Ceramic? But isn't that brittle?

Murr. When a high-speed steel bullet of over 2000 feet per second velocity hits a hard ceramic armour, a compressional shock wave travels forward in the ceramic and a similar shock wave travels backward in the bullet. These shock waves can reach pressures of 100,000 times atmospheric pressure and attain velocities of about 25,000 feet per second which is the sound velocity in the metal and ceramic. When the shock wave hits the rear of the bullet it is reflected as a rarefaction wave. The metal is stretched beyond its limit of strength and breaks apart — thus the armour has achieved its purpose, even though it may suffer a similar failure. However, you're not likely to be shot at with high velocity bullets . . .

Wells. Very well — it's yellow kevlar, then.

Murr. Or spider silk.

Wells. Spider silk? You're putting me on.

Murr. Not at all — spider silk is as strong as steel, weight for weight.

Wells. Don't tell me spiders make kevlar.

Murr. Not at all. Spiders are clever enough to convert all their waste to silk. They do not leave little white spots on the wall like flies. They produce a liquid-like organic polymer which is stored in a sac. To spin the silk it pulls it from the sac into long-chain polymers. It can

47

even vary the strength and tackiness, probably by varying the polymer chain length. If we cultivated huge colonies of spiders like the Chinese did with silkworms, we could produce smooth silk and weave it into clothing . . .

Wells. Brilliant — why there must be hundreds of billions of spiders around cluttering up ceilings and closets. If we organize them we'll revolutionize the world. You have my complete support. What do we feed the spiders?

Murr. Fruit flies.

Wells. How do we cultivate them?

Murr. We feed the flies rotten fruit.

Wells. Brilliant. *(Pause)* Excuse me, but does Professor Seitell often stand on his head like that?

Murr. Not when he's throwing teacups.

Wells. Good — keep him that way. Now in addition to a good bulletproof vest, I need a good warning system — in case intruders try to get at my magnificent art collection.

Murr. Ultrasonics.

Wells. I remember a little physics — that uses piezoelectric crystals — heh?

Murr. We've found a better way.

Wells. What's that?

Murr. Bats.

Wells. You're batty.

Murr. Not at all — bats avoid the need for piezoelectric crystals by having very fine cilia in their ears that vibrate at ultrasonic frequencies. The funny shaped ears focus the sound for them like an antenna. Bats are better than man-made devices because they're always moving around — 4π coverage.

Wells. You are an amazing woman — but, what do you feed the bats?

Murr. Simple — dead spiders. The lifetime of a spider is about 200,000 feet of silk.

Wells. Brilliant idea — beautiful organization — the spiders eat the flies, the bats eat the spiders — don't tell me that *we* eat . . .

Murr. No — but you could feed them to your lackeys.

Wells. When can you start work?

Murr. As soon as our book is finished.

Wells. You mean you and bat-man standing on his head are writing a book?

Murr. Yes, sir.

Wells. Very well — good luck. You'll have an unlimited budget for this project. Good day, Dr Murr.

Murr. Good day, sir.

Wells. Good day, Professor.

Seitell. Good day. *(Walks out on his hands, with* **Dr Murr** *holding his shoe laces)*

Wells. When this project is completed they can't deny me my knighthood any longer.

SCENE SIX

*(*Father Goldenestor *is on the phone to the* Bishop*)*

Bishop. It came in the mail today — but £240?

Father. Yes, the horse paid 4 to 1. After the acquittal the race committee reversed the disqualification.

Bishop. Is the work on the horses to continue?

Father. I think so.

Bishop. I see — you wouldn't happen to know when Reflection is racing again?

Father. No, but I'll let you know.

Bishop. It's just that I haven't lost my interest in physics.

Father. Of course.

Bishop. Besides, horses are more interesting than mice — and remember — old physicists never die they just lose their crystal balls — ha — ha — ha.

Father. *(Forced smile)* Good-bye, Bishop.

(Curtain)

ACT THREE

SCENE ONE

(The newly founded Wells Institute − "Physics for Fun and Profit" − self-contained on the Wells' yacht, Publish or Perish, *moored on the Thames outside King's College, London. The plaque in the main stateroom lists under the founder's name the Board of Directors −* **Professor Seitell, Professor Schrodenberg, Dr Murr, Dr Spiegelglass,** *and* **Father Goldenestor** *as extra-terrestrial consultant. At the bottom is the Wells' Institute crest − "Man does not live by Wave Functions Alone". In the laboratory on board are* **Professor Seitell, Dr Murr** *and others, working on the home-printed newspaper.)*

Alice. This has been more exciting than falling down the rabbit hole.
(Dr **Spiegelglass** *rushes in, holding a telegram)*
S'glass. An urgent message from Captain Wells − he's moved the demonstration ahead three days.
All. Impossible! How can we do it? He's a slave-driver!
Alice. When do we sail?
S'glass. Tomorrow. *(He goes to laboratory bench and examines the printer)*
Alice. Are there still any problems?
S'glass. Noise is the major one − we can't have a device in the home printing a newspaper at 3 a.m. unless it's quiet.
Alice. To a physicist what is noise?
S'glass. The ear is remarkably sensitive to sound waves, in fact if all the people in the world shouted for one second and you collected all that energy you'd just about have enough to boil a pot of water.
Alice. Gee!
S'glass. Any random change in air pressure reaching the ear is noise if the changes occur on a time scale between 10 and a few thousand cycles per second. For example, if you clap your hands you build up the air pressure between them in about 1/10th of a second, and this pressure wave is transmitted by the colliding molecules in the air to your ear.
Alice. What is the difference, then, between noise and music?
S'glass. They're both pressure waves but our ear and brain are trained to recognize certain frequency patterns as music − others as noise.
Mad Hatter. Nowadays you can't tell the difference.
S'glass. Sometimes that's true. If you go into a factory, gears meshing,

50

hammers striking, steam escaping all produce noise. Industrial noise is the prime cause of deafness in man.

Mad Hatter. I thought it was marriage.

S'glass. We've designed the *Gazette*-printer with nylon gears instead of metal ones because nylon is softer and doesn't build up the air pressure as rapidly. The cover is lined with foam to absorb internal noise.

Murr. What about the special rate from the telephone company?

S'glass. That needs a member's bill in Parliament — Lord Bury-Moore is taking care of that.

Alice. How long to transmit an entire newspaper?

S'glass. 20,000 words can be sent by telephone in about 30 seconds, if speeded up. This is stored on tape and can be printed in about 10 minutes. Each page falls out as in a Xerox machine. You pick up the pages and put them in this re-usable binder and voilà! A newspaper with your morning tea — no folding problem. For Captain Wells the saving in distribution costs is tremendous.

Murr. What are the latest estimates of printer cost?

S'glass. For mass production our cost is £38.

Caterpillar. Who pays that?

S'glass. That's Captain Wells' problem. Anyway, we've got to reach Scotland in two days. There are 2000 homes set up with printers for a trial demonstration.

Mad Hatter. If I know the Scots, there are 2000 bottles of whisky included with the demonstration.

SCENE TWO

*(*Publish or Perish *is at sea — the staff are working on the Murr Plan for Colonizing Spiders.)*

Alice. Any success in speeding up the spiders?

Murr. In this particular cage I fed them the drug mescaline along with the fruit flies. It didn't speed them up — they just spun crooked webs. Actually, if we solve the problem of successfully converting spider fluid into smooth silk we may ultimately try to synthesize the liquid. The main attraction with spiders is that rotten fruit and vegetables becomes our basic raw material, whereas in a synthetic operation the source of hydrocarbons can be costly.

Alice. Can you get more than one spider working on the same web?

Murr. Very interesting point. Spiders are basically loners but we do have a report on a species that is communal. We're waiting for a shipment.

Alice. Does the silk vary very much among spiders?

Murr. X-ray diffraction and chain length-determination are yet to be carried out, so we don't know. Another hope is that we can add certain elements along with the rotten fruit — like fluorine — that will be passed on through the flies and spiders to the silk.

Alice. Why fluorine?

Murr. Teflon is similar to polyethylene — they both have a carbon backbone, but in Teflon the side groups are fluorine instead of hydrogen. This makes for interesting changes in properties. It's still too early to know.

(The Mad Hatter runs in)

Mad Hatter. You've got bats in your belfry if you think this is going to work — someone's left the laboratory door open and 20 bats escaped.

Murr. Oh, dear.

(They move into the Bat Laboratory)

We've certainly had problems here. For one thing these dead spiders — most of them were eaten by other spiders. And the bats are not interested in dead spiders — they only go after things that move. We've even tried piling dead spiders on this table, turning on the fan and blowing them around the room. Somehow the bats know they're dead.

Mad Hatter. Damn clever — these Chinese!

Murr. We're trying to fill the room with all sorts of sounds like amplified recordings of spiders spinning webs — but we've got nowhere.

Alice. What *are* they eating?

Murr. Mice — our food bill is staggering.

Alice. Is it safe here? Don't bats get into people's hair?

Murr. An old wive's tale — the bats are not that stupid. Their ultrasonic echo-ranging system using both ears is quite sophisticated. We can't duplicate it, particularly if we match the weight of the bat.

Alice. How will the bat-alarm work?

Murr. Simple — we're tuned into their particular ultrasonic frequency. I'll say one thing — if the bats are disturbed the combination of the alarm going off and a room full of flying bats — that would scare off any intruder.

SCENE THREE

(Head-Ink Town Hall — secret conference room — all are present.
Captain Wells *addresses the group.)*

Wells. How I ever let myself into funding these hare-brained schemes is beyond me. I should have known you physicists are worse than useless. You've only wasted time and money.

(Everyone is shaking in their boots)

Father. But sir — we can explain.
Wells. You can explain? Why don't you get yourself a new line, Father? Don't you ever read the Bible?
Father. *(Subdued)* I'll try that, sir.
Wells. When I was in Naval Intelligence such bungling would have meant a court-martial. Do you understand that? *(Glares at* **Seitell***)*
Seitell. Y . . . Y . . . Y . . . yes.
Wells. Yes, what?
Seitell. Yes, your highness.
Wells. Stop standing on your head. *(*Seitell *sits)* Let's go over this fiasco. As for you, Dr Spiegelglass, I'd like to smash you to pieces if it didn't mean seven years' bad luck. Look at this. *(Pulls out newspaper)* You call this a newspaper?
S'glass. I can explain.
Wells. Explain? You, too? Father, get Spiegelglass a Bible also.
Father. Hail Mary, Mother of Grace . . .
Wells. Not now, Father. Look at this newspaper — 22 pages and nothing but pictures of bellybuttons.
S'glass. The interface, sir.
Wells. I said bellybuttons! Not faces.
Murr. And how many letters of complaint did you get?
Wells. That's even more disturbing — not a single one.
Murr. Perhaps you've found a way to revolutionize the newspaper industry.
Wells. Sure, if I gave every reader a bottle of Scotch with each paper — they wouldn't care what I printed.

53

Caterpillar. You have a point there, sir.

Wells. You — there — can't you see the no-smoking sign? What's your name, anyway?

Caterpillar. *(Puts away pipe)* Cat . . . cat . . . cat . . .

Wells. Stupid name Catcatcat. What do you do here, anyway?

Caterpillar. I . . . I . . . I . . . tes . . . tes . . . tes . . . I . . . I . . . I . . . eat . . . eat . . . I.

Wells. That's the first sensible thing I've heard. Let's get to the next point, Spiegelglass. How did you ever get the Prime Minister's private telephone number?

S'glass. It was a mistake — mis-dialled.

Wells. And what a mistake!

S'glass. But — she doesn't have one of our printers!

Wells. No — but she does have a tape recorder. MI5 were up all night deciphering the message and finally produced a hard copy of the newspaper.

S'glass. But what harm was done?

Wells. What harm? What harm? Five of those bellybuttons matched perfectly with those of their top secret agents. They're in a panic — I've been called to No. 10 tomorrow. How am I going to explain that? *(Glares at **Mad Hatter**)*

Mad Hatter. Couldn't you tell the Prime Minister you'll sew on new bellybuttons, or how about Navel Intelligence?

Wells. You're stupid — you know that? Now for the next screw-up — this bullet-proof vest.

Murr. Did it fail, sir?

Wells. No.

Mad Hatter. *(Aside)* Too bad.

Murr. What's wrong then, sir?

Wells. Wrong? Wrong? I've been picking flies out of it all day.

Murr. We're working on that — it's still a bit tacky.

Wells. And so is your effort. Furthermore, Dr Murr, this Fledermaus soup you make from dead bats — every time my lackeys eat it the dogs in the neighbourhood all howl.

Murr. Yes — something to do with the high-frequency bat sounds sir. We're working on that.

Wells. And that's not all — this bat-alarm was supposed to recognize intruders, wasn't it?

Murr. Yes, sir.

Wells. So how come one of my Rembrandts has been stolen?

S'glass. It wasn't stolen, I removed it. It wasn't a Rembrandt.

Wells. Are you batty? I paid 700,000 for it.

S'glass. When we set up the bat-alarm I asked a friend from the National Gallery to appraise your paintings — he became suspicious of the Rembrandt, and we later confirmed it as a forgery.

Wells. How did you do that?

S'glass. X-ray fluorescence — a microscopic amount of paint is removed with a hypodermic needle and placed in an X-ray beam. The X-rays knock out electrons from the atoms and when other electrons take their place, X-rays characteristic of each element present are emitted. The museum has done this for many Rembrandts so we know the pigments he used. Your Rembrandt was quite different. I removed it from your gallery as a favour.

Wells. 700,000 down the tube and you call that a favour?

S'glass. B . . . b . . . but.

Wells. Never mind.

S'glass. You bought that painting in Italy during the war, didn't you?

Wells. How did you know?

S'glass. We found another painting underneath the Rembrandt. It showed up under ultraviolet light since the pigments fluoresce and give out light.

Wells. What was the painting underneath?

S'glass. A picture of Mussolini.

Wells. *(Dumbfounded)*

S'glass. That was 700,000 lire, wasn't it?

Wells. Alright, alright. Now I'm getting worried. I've heard that the boys on Bleat Street are worried about our printer — they may be out to get me. Also the bookmakers at the track may have their sights on me — so keep on your toes.

S'glass. Yes, sir.

Wells. You too, Seitell! *(Seitell stands on toes)*

SCENE FOUR

*(*Publish or Perish *is moored behind King's College. A banner over the bridge reads "Bon Voyage — Spiegelglass–Bayha; Seitell–Murr". Inside the main stateroom a party is taking place. A huge wedding cake with portions already cut out is on the central table, some rice on the floor.* Professor Seitell *in morning coat and* Dr Murr *in wedding gown are the centre of attention as everyone is drinking champagne.* Dr Spiegelglass *and* Melanie *are similarly attired and we*

55

realize this has been a double wedding. **Dr Spiegelglass** *is on the platform.)*

S'glass. Before I propose a toast I want to thank everyone for this wonderful party — particularly Captain Wells in providing *Publish or Perish* for this double honeymoon voyage. He regrets he could not be here but a Buckingham Palace invitation can't be ignored. You all know what that means — it's as simple as *knight* and day *(laughter).* We are delighted that some of our pitfalls have been overcome — the bellybuttons have been covered up; the flies removed from Captain Wells' vest; and the rights to Fledermaus soup have been sold to McDonalds who are breaking a precedent by introducing soup. Reflection won the Gold Cup at Sun Down Park paying 3 to 1, so all is well at Head-Ink Town Hall.

As to Polly and Prof., we can only hope that every bun in the oven grows to full brilliance *(laughter).* We all know that Prof. is only a theoretician but we're certain our own Polly will teach him to fill the gap — although it may not be easy if he insists on standing on his head *(laughter).* We congratulate them on finishing their book — only half the number of pages. We're not certain how the Judges will take to this when they discover the type is half the size.
Mad Hatter. Give them a magnifying glass *(laughter).*
S'glass. If I can digress for a moment and say how happy I am that Melanie has made an honest man of me. When we first met in Las Vegas I had sworn that I would give up gambling for good. The only time I had gambled in my life was when I got married and I lost. But everyone deserves a second chance *(applause).* So I propose a toast to the Seitell–Murr bond — may all their wave functions be orthogonal, may they live in happiness and may . . .

(An **Inspector** *enters the stateroom and mounts the platform)*

Inspector. Just a moment! I am Inspector Sir Lock Homes — I have some distressing news for you. My men are waiting outside and I ask that you all remain where you are. I'm sorry to break up this wedding party but I'm afraid this honeymoon voyage must be postponed.
All. What is it — what happened?
Inspector. Captain Wells has been murdered!
All. *(Gasp)*
Inspector. We have the suspect under arrest.

S'glass. Who is it?
Inspector. Professor Stefan Boltzius!
All. That fiend! How could he?

SCENE FIVE

(The Prisoner Interrogation Room at Scotland Yard. At one end under intense lights sits **Professor Boltzius**. *In the darkness at the other end sits* **Inspector Homes** *and unbeknownst to* **Boltzius** *sit* **Alice, Caterpillar, Mad Hatter, Seitell, Murr, Schrodenberg** *and* **Goldenestor**.*)*

Inspector. Now, once more — what were you doing at Head-Ink Town Hall when the bat-alarm sounded?
Boltzius. Nothing — just scared with all those bats flying around.
Inspector. Why were you there?
Boltzius. I had written to Captain Wells and had a confirmation over the phone to see him at around 3 p.m.
Inspector. What for?
Boltzius. An idea for improving the sound of violins and guitars. I was looking for financial support through the Wells' Institute.
Inspector. And how were you going to make better violins, Mr Stradivarius?
Boltzius. Chladni patterns by laser interferometry.
Inspector. Don't give me any fancy language.
Boltzius. It's a technique that watches the soundboard vibrate when a string is bowed. There are very definite patterns for each note and for the precise way the string is bowed. The laser light is reflected off the soundboard and the interference pattern is recorded. By comparing the pattern of a newly made instrument with a fine Stradivarius, we can make adjustments to its soundboard and try to duplicate the Stradivarius. The same with a guitar.
Inspector. So you wanted everyone to play a Stradivarius?
Boltzius. Why not?
Mad Hatter. *(Whispers)* He should hear *me* play.
Inspector. *(Turns to the others and whispers)* Does this make sense to you physicists?
S'glass. *(Whispers)* It's quite possible.
Inspector. *(To* Boltzius*)* What happened then?
Boltzius. I arrived at around 2.45 and decided to look for Captain Wells

since his receptionist wasn't there. When I opened his office door the alarm went off and I saw he was ... he ... was ... oh ... it's awful, awful. And all those bats!

Inspector. Did you touch anything?

Boltzius. Not a thing.

Inspector. Then why did we find your broken fingernail on the floor near his body?

Boltzius. It isn't mine.

Inspector. Show us your hand.

Boltzius. *(Holds out right hand)* See — all nails are intact.

Inspector. No — I mean the left one.

Boltzius. I play the guitar — I must keep the nails trimmed on my left hand.

Inspector. *(To* **Spiegelglass**, *whispers)* What do you think?

S'glass. *(Whispers)* He could be telling the truth — ask him for a sample of his fingernails for comparison with the one you found.

Inspector. *(Whispers)* Can you tell — is it like a fingerprint?

S'glass. *(Whispers)* Yes. The X-ray diffraction patterns differ for each individual.

Inspector. *(To* **Boltzius***)* Will you give us a sample of each fingernail?

Boltzius. But I give a concert next week — I can't play without them.

Inspector. You don't expect to play in prison, do you?

Boltzius. I am innocent and I refuse to give up any fingernails.

Inspector. *(Whispers)* We're at an impasse.

Seitell. *(Whispers)* I don't like the guy — but I believe him.

Inspector. *(Whispers)* How could anyone kill Captain Wells without setting off the alarm — Boltzius must be guilty.

S'glass. What about a motive?

Inspector. He asked for money and was turned down — he gets mad and strangles Wells — breaks a fingernail.

Seitell. But that's circumstantial!

Inspector. Boltzius has a temper — don't forget the shouting match he had with you at your trial.

Seitell. But ... but ... that's all so ... so ... circumstantial!

Inspector. Maybe any one piece of evidence, but look — he sets off the bat-alarm; he refuses to give us a fingernail; he's turned down by Wells; he has a temper — it all fits. I recommend prosecution.

Seitell. That's unfair.

Inspector. We need a quick prosecution in a crime of this magnitude. You physicists haven't had a good trial since the Oppenheimer Case.

58

SCENE SIX

(The Criminal Court at Old Bayleaf – everyone is present – **Professor Boltzius** *is in the dock with* **Lord Bury-Moore** *defending him. The* **King** *ceremoniously enters as before and takes the bench.)*

Bailiff. All be seated.
King. Is the prosecution ready?
Queen. We are.
King. Is the defence?
Bury-Moore. We are.
King. Then proceed with the Crown *versus* Boltzius on the charge of wilful and most foul murder.
Queen. I call Inspector Sir Lock Homes. *(He takes the stand)* Sir Lock, will you examine and identify the Crown's evidence, item No. 1?
Sir Lock. Yes, that is the left hand middle fingernail that we found near the body of Captain Max Wells. It probably broke off the intruder's finger in the struggle when Captain Wells was strangled.
Queen. Have you identified its owner?
Sir Lock. No, ma'am. Since Professor Boltzius was found at the scene of the crime we later asked him to provide a sample of fingernail – but he refused. We discovered, though, that all fingernails were cut off his left hand. He told us he played the guitar – ha ha.
Queen. When the bat-alarm went off, how long did it take your constables to arrive?
Sir Lock. A few minutes.
Queen. What did they find?
Sir Lock. Captain Wells strangled and Professor Boltzius bending over him.
Queen. Thank you – your witness Lord Bury-Moore.

(The house lights dim and a spot falls on **Bury-Moore***)*

Bury-Moore. No questions. *(Applause, house lights up)*
Queen. I call Miss Sadlers. *(She takes the stand)* Are you private secretary to Captain Wells?
Sadlers. Naturally. I'm Wells' Sadlers.
Queen. Did Captain Wells ever communicate with Professor Boltzius?
Sadlers. Yes, he turned down his plan to make Stradivarii because he knew Professor Boltzius and Professor Seitell did not get on together.

Queen. On the day in question did Captain Wells have an appointment with Professor Boltzius?

Sadlers. I have no record of one.

Queen. Your witness, Lord Bury-Moore.

(The house lights dim, spot on **Bury-Moore***)*

Bury-Moore. No questions. *(Applause, house lights up)*

Queen. The prosecution rests.

King. Your case, Lord Bury-Moore.

Bury-Moore. I call Professor Boltzius.

(A murmur goes up from the audience as **Professor Boltzius** *carries a guitar case to the stand. He opens it, removes his guitar, waits for the house lights to dim and the spot to hit him. He plays 'Lagrima' by Tarrega. The audience is in tears and applauds.* **Boltzius** *returns the guitar to its case)*

No further questions. *(Audience applauds)*

King. Any questions, dear?

Queen. None. *(***Boltzius** *returns to the dock)*

Bury-Moore. *(With the spotlight on him, he rises)*

To be or not to be, that is the question *(Applause)*. Something is rotten in Denmark as indeed at Head-Ink Town Hall. There is only one person who was close enough to Captain Wells to know how to jam the bat-alarm and strangle him, leaving the incriminating fingernail. That person is in this very court. *(Murmur from audience)* Inspector, seal every exit.

Inspector. Done, m'Lord.

Bury-Moore. Since everyone's fingernails have their own X-ray diffraction pattern, the constables will circulate through the entire court with clippers collecting samples of the left hand middle finger . . .

(Suddenly all the lights go out – a shot rings out – women scream – the spotlight goes on and we see **Lord Bury-Moore** *lying on the floor clutching his chest. The* **Inspector** *goes to examine him as pandemonium breaks out.* **Bury-Moore** *whispers into the* **Inspector's** *ear)*

Inspector. *(Rises)* Everyone stay where they are – constables, guard the doors. *(The* **Inspector** *helps* **Bury-Moore** *to his feet – the audience* *applauds as* **Bury-Moore** *raises his hand for silence)*

Bury-Moore. *(Opens his coat)* Thanks to Captain Wells' spider-silk
bullet-proof vest which he no longer needed and which I am wearing, folded over double, of course, the strangler of Captain Wells
has just revealed his hand. The very hand that fired that shot will be
found to have a broken fingernail to match the one found at the
scene of this most vile crime. And that man is . . .

(The **King** *stands up and aims a gun with his left hand at* **Bury-Moore**
but the **Inspector** *is alerted and grapples with him, removing the
gun)*

Mad Hatter. I've read plenty of Agatha Christie but this is the first time
it's been the judge!

Bury-Moore. I call the King to the stand. *(As the* **King** *is led to the stand
by the* **Inspector,** **Bury-Moore** *pulls off the King's judicial wig to
reveal a completely bald pate.* **Bury-Moore** *dons the wig himself)*
This may be a bit irregular but I shall serve in the dual role of judge
and defence attorney. *(Applause)*

Mad Hatter. If this doesn't get him an Academy Award nothing will.

Bury-Moore. Your name?

King. *(Silent)*

Bury-Moore. There are ways to make you talk.

King. I am King Louis of Carrollina.

Bury-Moore. A country which does not exist!

King. It does in the hearts and minds of those who love her.

Bury-Moore. What a corny line — who writes your script?

King. *(Silent)*

Bury-Moore. Tell us why you strangled Captain Wells.

King. Alright, alright, I did it! *(Hushed silence as the spotlight changes
to the appropriate colour for courtroom confessions)*

SCENE SEVEN

King. You are right — Captain Wells and I go back to the war. We were
both in Naval Intelligence. I was 006 and he was 007. He always was
one-up on me. We were working on the German submarine menace
and came up with the idea of using trained dolphins with strong
magnets attached to their snouts. Everyone should have a porpoise
in life. Ours would wait outside the submarine base in France and

attach themselves rigidly to the bottom of the submarine. In this way we thought they would not be detected when the submarine surfaced. *(Pause)* Unfortunately we forgot that dolphins breathe air and lost the first few. *(Pause)* We then shifted to a small magnetic loop attached to the dolphin's belly which would pick up the residual magnetism in the submarine's hull — activate the loop and tickle the dolphin's belly. The dolphins were trained to follow this stimulus. An antenna on the dolphin's back would radio a signal back to Portsmouth. *(Pause)* Unfortunately, the Germans thought it odd to see a dolphin with an antenna and shot them. *(Pause)* We then devised a loop antenna surgically implanted under the dolphin's skin. This transmitted beautifully. *(Pause)* Of course, the Germans also heard the transmission since they were a lot closer to the dolphins. We were now panic stricken — it was 1942 and the Germans were developing a liking for dolphin meat — we later captured one of their cook books — but the loss of allied shipping was prohibitive. I — mind you *I* — then devised an antenna that would only transmit when the German submarine was submerged so they couldn't pick up our signal — you see salt water absorbs all radio signals. *(Pause, pause, pause)*

Bury-Moore. And?

King. It finally worked and we won the Battle of the Atlantic. *(Applause)* That operation has remained secret all these years — first for military security and then to prevent Greenpeace from annoying the Ministry of Defence. Finally, the Queen decided to reward Captain Wells with a knighthood. He was 007, I was 006, naturally he got the credit. I begged him to tell the truth about my role but he refused. I decided to go to Head-Ink Town Hall.

Bury-Moore. But the bat-alarm!

King. Yes, I was a frequent visitor and knew how to deactivate it. A tape filled the room with the same frequency ultrasound as the bats generated — this confused them. I begged the Captain for the last time — he was about to leave for the Palace. I became enraged and . . . and . . . you know the rest. I hate him!

Queen. I can hardly wait to prosecute him!

Bury-Moore. Inspector — take him away. *(Raps gavel)* Professor Boltzius stands acquitted! *(Applause)*

(Curtain)

ACT FOUR

(On board Publish or Perish. *A second bon-voyage party with every-one present including* **Sir George Inversion**, *Head of the Weather Bureau who is addressing the group.)*

Sir George. I am delighted to wish bon voyage to the *former* newlyweds and I've prepared a special one month long-range weather forecast especially for the journey. *(Gives to the two couples)* Compliments of Her Majesty's Weather Bureau.

S'glass. How can you predict weather so far in advance?

Sir George. It's not easy — we need huge computing capacity. Basically we divide the world into a three-dimensional matrix, each cube approximately ¼ mile on a side. At any time the air in each cube has a certain pressure, temperature, velocity pattern, moisture content etc. — all this is matched at each cube boundary. We take account of the energy from the sun entering from above and absorbed by the ocean and land below, as well as the energy reflected and trans-mitted by clouds, ocean, land mass etc. In principle, if done care-fully we can predict weather for four days.

S'glass. But you said two months . . .

Sir George. You'll be on your honeymoon then — you'll hardly notice.

S'glass. How large is the computer?

Sir George. The largest in the world.

(Enter **Boltzius** *and* **Bury-Moore** — *everyone applauds)*

Boltzius. I'm here to congratulate the newlyweds and to thank Professor Seitell who believed in my innocence and obtained the invaluable help of Lord Bury-Moore.

Seitell. *(Standing on his head)*

Murr. Up boy, up! *(Seitell stands erect)*

Seitell. When I was in jail awaiting the *(coughs)* 100,000 ohms I was forced to read Professor Boltzius's book. I realized he had a lot to say. And when I learned he played the guitar, I knew he couldn't be all bad. So I convinced Lord Bury-Moore, who had collected many a fee from Captain Wells, to defend Professor Boltzius as a final tribute to the Captain. *(Applause.* **Spiegelglass** *takes the platform)*

S'glass. As you know, the untimely death of Captain *Sir* Max Wells means we must make the Wells' Institute self-supporting and run a tight ship. Our motto "Physics for Fun and Profit" will be changed to "Physics for Profit" — the Fun will go undercover.

Mad Hatter. Sounds like a newlywed talking. *(Laughter)*

S'glass. We shall have to examine all programmes most carefully. You've all been asked to prepare brief proposals. These will be appropriately bound and interred with Captain Wells who will read them at his leisure. His final request to be buried at sea was made just as that dastard King Louis strangled him. *(Hissing from everyone)* As a closing tribute we shall have the house lights dimmed — with an appropriate spotlight for research proposals, *and* with Professor Boltzius playing his guitar offstage we'll hear from each of you. First, Mr Caterpillar.

(Caterpillar takes stage without hookah)

Caterpillar. As you see, I have given up smoking. *(Loud chorus of approval)* I have discussed with the Director at 'Q' Gardens the problems of pests that eat leaves and trees. He advised me that certain flora in Australia have high concentrations of fluorine and the leaves are toxic to the fauna. Our plan is to feed soluble fluorine to young saplings and follow the fluorine uptake with an NMR coil round the trunk — this is very sensitive since the fluorine nucleus is very magnetic. It may take a few generations of trees to improve the species since many will die. We hope the fluorine will strengthen the wood and make it more fire-retardant. *(Applause)*

Mad Hatter. A lot of caterpillars may go hungry.

Caterpillar. Who needs caterpillars? They're so squishy when you step on them.

S'glass. Since Mr Hatter is anxious to be heard, he's next.

Mad Hatter. We shall develop unbreakable ceramic tea sets but with an added effort to help artists. Most pigments used by potters are fairly standard. They are painted on and fired. We plan to use new minerals and optical analysis to create beautiful glazes. We want to know which electrons produce these colours and we will search for new, cheap minerals to expand the range of colours. We shall even engage a few theoreticians to calculate the appropriate wave functions and energies. *(Seitell stands on his head)*

Murr. Up boy — up! *(Seitell returns to feet — applause)*

S'glass. We all must agree that Mr Hatter is not so mad in trying to combine physics and art — "An unbreakable thing of beauty is a joy forever." *(Applause)* Father, you're next.

Father. Our parish was so successful in betting on Reflection that they decided to engage a less photogenic Minister. *(Shrugs)* This left me temporarily at liberty. I'm glad that my sports project has been approved. I intend to divert my attention from horses to athletes.

Mad Hatter. Quitting when you're ahead?

Father. Actually, while the Bishop is ahead. I shall develop ultrasonic scanning and pressure transducer measurements to determine the strain athletes place on their bones and muscles under stress of sports. We shall try to develop exercises to strengthen these parts as well as special stress-absorbing clothing. Since insurance companies are often required to pay large sums for sports injuries they are partially supporting this work. *(Applause)*

Mad Hatter. I suppose you hope to jog into Heaven?

Father. No — I'm going to pole-vault over the Pearly Gates.

S'glass. Alice?

Alice. We're going to make clothing out of spider silk. We have to learn which chemicals to feed spiders so as to colour the silk. This complements the spectral analysis work of Mr Hatter. We don't think the world is ready for transparent clothing — especially when you're only 13.

Melanie. I wouldn't mind transparent . . .

S'glass. . . . Stow it Melanie. We are pleased to announce that Lord Bury-Moore has agreed to join us. *(Applause)* He will now tell us his exciting project.

Bury-Moore. Let me make it clear — I will *not* be studying the role of physics in Shakespeare's plays! *(Laughter)* My subject is oral communication. Physicists must be taught how to apply the fundamentals of good theatricality to their talks. After all — why spend £20,000 on a piece of research if it is not effectively communicated to your audience. We must study the type and style of talk that will be retained by an audience and train speakers to recognize good theatricality. This subject cannot be learned from a textbook. Just like dancing, you must do it to learn how.

Seitell. I second that!

Murr. Down boy, down!

Bury-Moore. We shall work with groups of scientists and evaluate the effectiveness of talks, not only immediately but even months later. Thank you. *(Applause)*

S'glass. If every scientific talk ends with a shot in the dark — we won't mind transparent spider silk. *(Laughter)* Now for the Murr–Seitell superconducting pairs we must defer to the distaff member. *(Applause)*

65

Murr. The structural integrity of airplane frames is of vital concern and the periodic X-ray and ultrasonic testing for cracks and defects is very costly. We plan to add optical fibres to the carbon fibre–epoxy composites so that any crack will show up as a break in optical continuity. In this way on-line testing of an entire aircraft, even in flight, can be performed with a computer monitoring all stress levels. Even a small stretching of an optical fibre will alter its light-conducting properties. *(Applause)*

S'glass. Excellent idea! And so it falls to me to make the final presentation.

Melanie. Hold on – aren't we married?

S'glass. You're not starting already?

Melanie. Didn't you just say the distressed member?

S'glass. I said *distaff* member.

Mad Hatter. Is there a difference?

S'glass. For the final . . .

All. Let Melanie speak! Yes!

S'glass. *(Throws up hands in despair)*

Melanie. Just because I studied in Las Vegas doesn't mean I can't be interested in physics.

S'glass. This I got to hear.

Melanie. Now, admit it, Spiegelglass, wasn't your research proposal my idea?

Mad Hatter. Also the marriage proposal. *(Laughter)*

S'glass. You're both right. *(Laughter)*

Melanie. *(To audience)* You see – one day I got stuck with this cheap face powder – the colour was terrible.

S'glass. C'mon dear, no one looks at your face.

Melanie. I do! So I said – "aren't you scientists supposed to know about colour?" No answer! So I pulled the plug out of his computer. "What happened", he says – so I held the plug in my hand and said – "You know what I'm going to do with this plug?" Now listen! I want to go into a beauty parlour, look at a colour chart and say "that's the colour I want – measure my face colour and tell me what make-up to use." Simple, heh? Even Spiegelglass admitted that.

S'glass. Yes, dear.

Melanie. So, now he's going to invent a machine that even beauty parlours can afford. *(Applause)*

S'glass. Melanie – that was excellent!

Melanie. So now, Spiegelglass, answer me one thing.

S'glass. Yes, dear.
Melanie. Tell me what a wave function is!

(As Publish or Perish *sails off into the blue with* **Professor Boltzius**
still playing the guitar, the 'almost' newlyweds wave from the stern
to those on shore. We see that the bon voyage banner has appended
to it a second banner reading "We've all studied a little physics."
Alas! All stories must end but it is hoped everyone will visit Wonder-
land again, page 1 – now what was that question about wave
functions?)

He thought he saw a wave function
With real asymmetry
He looked again and thought
Perhaps – orthogonality?

(Curtain)

67

Captain Wells' Appendix

Two days after burial at sea, Captain Wells' last will was probated in London. One of his vital organs (the appendix) was left to the Wells' Institute. This provision in his will was telegraphed to *Publish or Perish* which reversed course to the burial spot. Fortunately, Captain Wells was old enough to still be floating. His appendix was removed by Dr Spiegelglass, lead weights attached and the body heaved over the POSH side. Surgically appended to his appendix was a microchip which was duly read by the ship's computer:

I, Captain Sir Max Wells, being of sound mind and corpulent body do hereby bequeath this microchip to the Wells' Institute. Firmly in the belief that Physics is too important to be left to scientists, I leave to the Institute these 10 commandments which damned well better be followed by staff.

1. Read the *New Scientist* – better yet, obtain your own subscription.
2. Get a job as a consultant. This is the second oldest profession, different from the oldest only as to whether you lean over forward or backward to please your client.
3. Reread the *New Scientist*, especially the advertisements – you may not be with us long.
4. Do not employ mathematics until you understand what is happening. Do not employ mathematics after you understand what is happening – let the accountants do that.
5. If you come up with a money-making idea, you really don't understand the situation. (Would you mortgage your home on the idea?) Money and angular momentum (in that order) make the world go round.
6. If you build a better mousetrap you may have a saleable item but remember – the mice and the competition are getting smarter.
7. Let the PR guys sell it first – then invent it.
8. Money is to our world what wave functions are to electrons. They both provide restrictions on the most efficient stable configuration.

9. If you feel secure in your job because I'm buried at sea — remember there's always another **SOB** ready to take my place.
10. All subversive literature such as *20 Ways to Show You're a Physicist* is banned from the Institute. Now — get to work.

A messenger was sent to the King's College Physics Library where the banned item was found as part of the student reading list:

20 Ways to Show You're a Physicist

1. Never answer your mail. 99 per cent of those who write to you are in an inferior position. Why else would they write? However, requests for reprints of your articles should be answered immediately to ensure rapid propagation of your brilliance.
2. When reviewing a book be certain to find at least four errors in judgement or fact. Remember, as a book reviewer your superior position has already been acknowledged — it is your task to justify that trust.
3. Never address an audience without a piece of chalk in hand. The listener's eyes are glued to this weapon waiting for it to strike.
4. Never give a talk without a mathematical derivation.
5. If anyone in your audience is not taking notes, glare at him unmercifully. The gems of your oral utterances deserve to be recorded for posterity.
6. Learn the music of at least half a dozen early composers, preferably all before Bach.
7. If athletic, tennis is your game. It provides the most frequent interchange of energy and momentum.
8. Under no circumstances dress like a businessman. You are a member of a select group — dress with an appropriate disregard for convention.
9. If ever you appear at work before 10 a.m., your excuse should be that you worked on an experiment all night.
10. When sitting in committee, find fault with all suggestions. After all, ideas are fallible — make that point.
11. If you don't wear glasses, cover up this inadequacy with an even greater disregard for conventional dress.
12. Develop a taste for Szechuan or Indonesian cooking — a mark of discrimination and worldliness.

13. When delivering a paper at a meeting, snow your audience under with undecipherable slides. A well-designed slide is equally mystifying independent of how the projectionist inserts it.
14. Smoking is out, marriage and children are in.
15. Leave your office blackboard filled with undecipherable mathematics. A "do not erase" sign should be prominently displayed.
16. Always leave a copy of the *Physical Review* on your desk, opened at an article filled with equations.
17. Do not drink the hard stuff. Become familiar with the names of obscure California vineyards.
18. Do be a name-dropper. Leave your desk calendar open, indicating a future date with some well-known personage.
19. A neat, tidy desk is a reflection of a blank mind. Periodically move the mess around.
20. When people ring, be certain your secretary always says you're out. Return only 5 per cent of your calls. Leave a note on your desk signed by your secretary that an important person called *twice*.

The key to proper behaviour is arrogance — God's way of saying he didn't make us all physicists. It is a small price for the public to pay for the atomic bomb and television.